Other Books on

VEDIC MATHEMATICS SERIES

Vedic Mathematics New Horizons
Initial Lessons (New) 175/-

Vedic Mathematics New Horizons
Advance Lessons (New) 225/-

3. Vedic Mathematics for All 175/-

. Learn and Teach Vedic Mathematics 150/-

. Practice Vedic Mathematics 195/-

6. Vedic Mathematics Decodes Space Book 195/-

7. Vedic Mathematics Basics 125/-

8. Vedic Mathematics Skills 150/-

9. The Teaching of Vedic Mathematics 125/-

10. Learning Vedic Mathematics on First Principles 125/-

11. Vedic Geometry Course 195/-

Unit No. 220, Second Floor, 4735/22,
Prakash Deep Building, Ansari Road, Darya Ganj,
New Delhi - 110002, Ph.: 32903912, 23280047, 09811594448
E-mail: lotuspress1984@gmail.com, www.lotuspress.co.in

Vedic Mathematics

Vedic Mathematics Skills

Dr. S.K. Kapoor

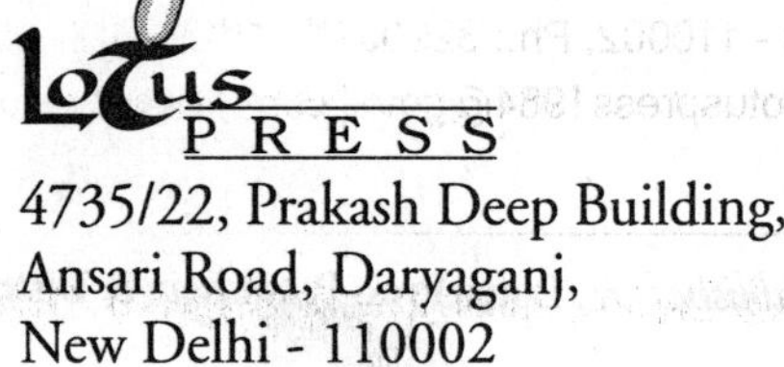

4735/22, Prakash Deep Building,
Ansari Road, Daryaganj,
New Delhi - 110002

Lotus Press Publishers & Distributors
Unit No.220, 2nd Floor, 4735/22, Prakash Deep Building,
Ansari Road, Daryaganj, New Delhi- 110002
Ph:- 32903912, 23280047, 098118-38000
Email : lotuspress1984@gmail.com
Visit us : www.lotuspress.co.in

Vedic Mathematics Skills (H/B)

ISBN : 81-8382-059-X

Printed & Published by : **Lotus Press Publishers & Distributors,** New Delhi- 2

INNER VOICE FOREWORD

— FROM A SAINT

Hear your inner voice and be free from the outside limitations. Within is the seat of the Lord and it is here where we have to be face to face with our selves. This origin source reservoir is to be tapped, and this appears to be the urge and intensity of present attempt to reach at 'Vedic mathematics basics' in five steps. These five steps would prepare for the phase and stage to be in communication with the SELF.

The SELF is TRANSCENDENTAL. The seat of the SELF within *Triloki* is the Creator's space. Self established at the seat of Creator's space ensures carrying the Being along the transcendental carriers. It is the bliss of this ensured supply of transcendental carriers which intensifies the urge of the transcending mind to glimpse the inner folds of the transcendental worlds manifesting within Creator's space.

The whole efforts and beginning for it, as such is to be to attain transition from Triloki manifesting as outside mundane macro state world as first state of consciousness for the intelligence to the next transcendental micro state world of the intelligence lively within Creator's space as origin source reservoir of Triloki. Our ghan / body cube, as such, is to be chased for its origin source reservoir within it lively as Chatar mukhi / four head Creator.

Really blissful is the learning path of this chase range. It is transcendentally blessed. Let sadkhas permit their transcending mind to fulfill with bliss of the transcendental worlds / Panch Mukhi / five head lord sustaining the Creator's space.

Lively within
cavity of heart:
25-11-2005

In search of ***JYOTI***
manifesting as **NAD**

■

PREFACE

Urge to know ancient wisdom expressed as Vedic knowledge inherently imples acceptance for chase of it the way these systems permit. This as such would mean that the sadkhas for the chase of Vedic systems have to proceed along the established paths of *Sankhiya Nishshta* and *Yoga nishta* which not only run parallel to each but are also complementary and supplementary of each other.

The sankhiya nishtha, on its ultimate analysis amounts to availing the artifices of numbers, for coordination arrangement of organisation format, while on the other hand, the Yoga nishta avails the dimensional formats, for regulation of transcendence and ascendance of mind through the manifested creations. The **TRANS**, also means, the number ahead, and it is this way, the transcending mind ultimately avails the artifices of numbers to sequentially glimpse the ultimate core of the transcendental worlds.

The par-excellence feature of Vedic systems is that these approach the whole range of knowledge as a single Discipline, and as such, on ultimate analysis the organisation of Discipline of knowledge avails the dimensional frames parallel to the artifices of numbers. This, this way, shall be taking us to the artifices of numbers, and the dimensional frames as the source processing tools in terms of which even the self referral feature of transcending mind chasing its own transcending path can also be worked out intellectually.

It is this feature which deserves to be chased for the basics, and here in the present attempt, the urge being to learn and teach Vedic mathematics on its basics, as such, at this phase and stage of exposing young minds during Middle school learning, The focus is to be upon the centre of the cube / origin of 3-space being of the format of hyper cube-4 and as such seat of creator's space.

This focus naturally aims to eye the goals of attainments for the young minds for perfection of their intelligence by having transition from the set up of the Cube to transcend ahead for glimpsing the set up of hyper cube-4 along with its transcendental base of the format of hyper cube-5.

Though, the learning all about hyper cubes 4 and 5 is to be the content focus of high school classes, and ahead there to would be the subject focus of the formats of hyper cubes-6 and 7 during higher secondary classes, but the initiations for the foundations of appropriate mental state for those formats are to be ensured during schooling of middle school stage.

This way the learning range of this phase and stage of schooling during classes 5 to 8 is to be centred around the set up of the cube with focus upon its centre /

origin of 3-space as seat of creator's space (4-space) as of format of hyper cube -4. Accordingly the teaching and learning of this range of schooling is phased as five books namely, book-1The teaching of Vedic mathematics, book-2 Learning Vedic mathematics on first principles, book-3 Vedic mathematics basics, book-4 Vedic mathematics skills and book-5 Vedic geometry course, for perfection of intelligence.

These books, in their integrated approach of chase of the range for set up of the cube / 3-space with focus upon its center / origin of 3-space, as source reservoir for the features of the set up, is meant for exposure of young minds to acquire perfection of intelligence while learning basis of basics on first principles along the concrete format of a cube.

The first principles for approaching basis of basics of Vedic mathematics on solid formats are being supplied by the Ganita Sutras, particularly as the ordering principle of Ganita Sutra-1 which makes it as the source Sutra, and as such 'cube' and 'Ganita Sutra-1' are to ever remain within the eye sight of the learners of this phase and stage of schooling.

The aim and goal of learning being the chase of basis of basics on first principles to ensure perfection of intelligence for the young minds, as such heavy responsibiliy lies upon the Vedic mathematics teachers and parents to make the learning really blissful throughout the duration of chase.

The whole range of enlighened souls begining with Maharishi Valmiki and Maharishi Ved Vyas to Maharishi Dayanand and Swami Bharti Krsna TirthaJi Maharaj to Shri Pad Baba and Maharishi Mahesh Yogi, this learning path is being made blissfully lively, and the enlightened teachers are to continue the linege.

With these inspired background, present humble attempt has been made to reach at these books and it is hoped that these may provide required exposure for the young minds with the help of the enlightened teachers, who shall be making their instructions free of technicalities and the knowlede content to flow blissfully within the head s and hearts of the students.

Mr. Kenneth R. Williams and Mr. R. P. Cosmic Kapoor deserve my heartful thanks for remaining intellectually with me during these days of my reaching at these books. Mr. Rohtash Kapoor deserves full appreciation for conducting Vedic geometry course.

Mr. Deepak Girdhar deserves all appreciation for his nice graphics and the computer work.

Credit for nice publication goes to Lotus Press.

The inner voice forward from a saint gives me the satisfaction and confidence as that these efforts may be fruitful for the cause of basics to be brought home to the young minds.

Vedic home, **Author**
Rohini, Delhi
25-11-2005

■

CONTENTS

■

SECTION - I

BASIC OPERATION REVISITED

1

CONCEPTUAL FOUNDATION

1.1CONCEPTUAL FOUNDATION

CONCEPTUAL CHASE

With an aim to lay a proper foundation for learning 'arithmetic to astronomy' along the format of ancient wisdom of Vedic mathematics systems in terms of the first principles of functional rule of Ganita Sutras, beginning is to be had by revisiting the way the basic operations of arithmetic need be approached.

The basic operations of arithmetic are four, namely, addition, subtraction, multiplication and division. These are also interrelated and on first principles have inherent sequence. The first amongst them, with which the beginning is to be had, is 'addition'. The addition and subtraction are interrelated which relations ship may better be appreciated in terms of the pair of opposite orientations permissible along the set up of a line. While the addition operation may be accepted as a ***Vridhi*** / increase along, say an orientation from 'right to left', the subtraction in the process is taken accepting the opposite orientations from left to right, and the same in reference to the increasing process of addition, it emerges as a decreasing process.

Further, along the format of a line of increasing and decreasing orientations, the first, namely, the increasing orientation attainments as step by step addition, as well is to be viewed, when a particular number of steps of addition are to be taken in consideration, as a distinct feature of addition operation, and it, as such, emerges to be a multiplication operation along the format of a line. Likewise, the subtraction operation, results into and gets interconnected with the subtraction operation.

This way, all these four basic operations namely addition, subtraction, multiplication and division operation get inherently inter related as of the same generic whose glimpsing is to provide the needed insight for perfection of intelligence. It is only in terms of glimpsing of the common generic binding all the basic operations that smooth transition can be had for these operations from the format of a **line** to that of a **spatial order of plane**.

Primarily it is for this aim of attaining desired transition for the initial format of a line for the basic operation to have their applications along the spatial order of the plane, that these operations deserve to be revisited in terms of the first principles of the functional rules of Ganita Sutras, which inherently being of self referral features unfolding in a sequential order providing required transition from linear to spatial, and from spatial to solid and further from solid to hyper solid formats for the numerals of place value systems. It is this feature of the Vedic mathematics systems which deserves to be very gently chased by the young minds under the able guidance of the enlightened teachers who having already visited this feature of approach to the artifices of numbers.

Here at this phase and stage of learning of this delicate feature of artifices of numbers for transition from the format of a line for their expressions to that of spatial order expression domain of plane, opportunity may be availed by the teachers to impressed upon the young minds as that the format of 'line' is of a linear order, as much as that it is just 'one axes', while as that the expression domain of plane, is to require the availability of a pair of axis, and ahead the expression domain of solids would be requiring availability of triple axis and further ahead the hyper solid domains requirement would be for availability of more and more number of axis.

Such being the need of the expressions domains, which at the initial stage of the format of a line, being of only single axis, as such the numerals '1, 2, 3, 4, 5, 6, 7, 8, 9', could be expressed as 'single digits'. However when there is to be a transition from the format of a 'line', to that of a plane, then, naturally there would be a requirement for availability of 'double digit' expressions for numerals as well. This has such, provides us a shift from single digit numerals '1, 2, 3, 4, 5, 6, 7, 8, 9' to double digit numerals '01, 02, 03, 04, 05, 06, 07, 08, 09'.

Here at this phase and stage of learning, the teachers may avail an opportunity of impressing upon the young minds as that, this wonderful attainment of transition stands attained only because of the availability of the tenth numeral '0' as of place value features. And that, it is, by its appropriate utility of application for a shift from single digit expressions for numerals to that of double digit expressions for them that the needed transition stands successfully attained.

Here itself, further opportunity may be availed by the teachers to impress upon the young mind as that the place value numeral '0', is of general features, and that the same is not restricted to ten place value system, as that, it is to play its similar role for all other place value systems as well.

The transition attained from single digit expressions format of a line to that of double digit expressions for the numerals, when chased shall be bringing us face to face with the widening gaps for the expressions formats of 'addition' and 'multiplication operations', on the one hand, and of addition and subtraction operations on the other hand. These distinguishing features of addition and multiplication along a format of a line in terms of single digit numerals and along the format of a plane in terms of double digits numerals deserves to be chased very gently as, in fact, it is needless to impress more, as that the conceptual seed of 'generic of operations' is lively at this junctions point being the source reservoir.

With availability of the plane format for expression for basic operations, the addition operation, as is evident from manifestation of its symbol (+) as a pair of axis constituting a dimensional frame, which, in a way amount to providing a boundary for a plane. On the other hand the multiplication operation as it evident from the manifestation of its symbol (x), it amounts to a angular shift for the axis resulting into a shift of the order which amounts a shift from the boundary of a plane to the domain of the planc itself.

This, as such distinguishes the addition operation as a vridhi / increase along the boundary of a plane / square / circle, and as such its format essentially remains to be that of a line / 1-space, while the multiplication operation gets it fulfillment with availability for its complete expressions within the spatial domain of a plane. The multiplication operation marks its distinctguishing feature over the 'addition operation' as that 'multiplication operation' take to 2-space / surface / plane / square / pair of axis / double digit expression for numerals and numbers. On the other hand the addition operation sticks to its original features of line format for its exprsession, though with the shift from single digit expression to double digit expressions, it remains satisfied with the role of line / 1-space as domain in its original format of single digit numerals stage to that of the role of a boundary of 2-space / surface / plane / square with availability of numerals of double digit expressions.

Now with the shift from single digit formats of numerals to that of double digit format for numerals, there is a shift from a line format to plane format, and with it, the linear expression order for numbers gets transformation into a place value expressions format for the numbers. As such, the addition operation and multiplication operation with their double digit numerals, though avail the common format which comes to be that of a 'cross' / 'manifested format of multiplication', but the operations get split into steps of pairing.

The pre-requisites for approaching, addition and multiplication operation on first principles of 'pairing' along the format of 'cross' are

(1) as that the numerals 'zero to nine' are to be approached as 'picks', as 'number of times' or simply as 'times'. The 'zero time' pick yields 'zero value' / 'zero number (0). Then, 'one time', as pick of 'one' is to yield 'one' unit value / number 'one' (1).

(2) Further, as that the pairing along the format of a cross is to be as follows of (i) pairing of ends on sides (ii) pairing of ends crosswise, which shall be a pair of steps and (iii) the pairing of ends of the other sides.

This set of four pairings of ends of the cross, are to be taken as four pairings of four digits of two numbers of two digits each. The joint of the cross (of pair of axis) provides a joint for the middle pair of pairings. This way the set of four pairs get reduced into three values, firstly as of pair pairing ands (A, X) on side, secondly the middle pair of pairings of ends cross wise [(A, Y) & (B, X)] & thirdly the pair of pairing of ends (B, Y) on the other side.

Now when the operation is being conducted at first place of the place value system, it being first, and hence of linear format, as such it is carrying features of addition, and accordingly all the above three values are added, as linear addition operation.

Further when the pair of place values are permitted to be availed, with first and second places being of distinct values of the order $10^0 = 01$ and $10^1 = 10$, with

01 and 10 being a reflection pair, and crosswise opposites, so here in such a situation, the mirror in between the reflection pair of artifices (01, 10) remains un manifest, because of which, both digits of a numerals or of a double digit number, gets placements side by side impliedly indicating that they are glued because of the format beneath which is a pair of places of place value system. With it, the multiplication operation, as pairing of digits with placements along ends of cross (lines), and as such, the three values, get side by side placements in the sequence and order of their emergence, and same to be taken as glued in terms of the place values of the format beneath, and their by the multiplication operation completes its attainment.

Here, at this phase and stage of learning, the teachers may avail the opportunity to impress upon the young minds as to the distintguishing features of addition and multiplication operation. The focus need be concentrating upon the feature of a transition for multiplication as repeated addition along the format of a line, to addition and multiplication expressing their full features along different formats, that is along a line / boundary and along a domain of a plane format.

Here further, the opportunity may be availed to impress upon the feature of repetitive multiplication operations as 'exponents' / powers / degrees as 'Gunas' range, as comparison to the feature of repetitive addition operations as 'multiplication'.

Here further, the opportunity may be availed to impress upon the feature as that as along the format of a line, the multiplication operation was getting superimposed simply as a repetitive addition, and likewise the repetitive multiplication, along the plane format shall be expressing as superimposed 'exponents' as multiplication along a format of a cross.

Here further, the opportunity may be availed to impress upon the young minds as that, as for full expression of multiplication transition is being had from the format of a line (where the multiplication as repetitive addition get superimposed as addition along the line) to the format of a plane, (where only the multiplication attained the format of a cross within the domain of the surface), so likewise for full expression of 'exponents', transition is to be had from the format of a plane / square / 2-space to the format of a solid / cube / 3-space.

This attainment for full expression of 'exponents', (which otherwise as repetitive multiplication stands superimposed upon multiplication of cross format within a plane), there is a need for a solid format of cube / 3-space for which the numerals have to have a transition for their double digit formats to triple digit formats.

As such the necessary focus and attention is to be upon the proper initiatives for full exposure for the young minds for comprehension and insight as to how the features of different basic operations unfold differently along different formats and that transition from one format to another would require a perfection of intelligence. For it, the students are impressed upon to chase the multiplication operation, the Vedic way, for its take off from the phase and stage of 'repetitive addition from the format of a line as expression domain for single digit expressions only to, as a first step to that of multiplication of the format of a cross within the domain of a surface, and then to the format of solids to free the exponents from the feature of simply being repetitive multiplications to their full expressions within the solid order which is to lead to the transcendental worlds as a phenomenon of transcendence and ascendance. This in fact is going to be the phenomenon of the transcendental core of the Creator's space (4-space of spatial order).

With it the exponents operation would lead to a transcendence process at center of the grids sustaining the surface plates of the cube / solids. With it a phase and stage would emerge for transition from plane grids to solid cages, for sustaining solid contents lumps, whose static positions would manifest solid boundary of the Creator's space while whose flow would lead to the transcendental worlds.

A step ahead would be awaiting the transcendental carriers of hyper solid order which shall be leading the Creator's content to self referral domains lively within the orb of the Sun of features of Vishnu lok / 6-space as existence domain of Divya Pursha / atman, as is the enlightenment of senior sadkhas preserved in the available scriptures.

As such, the sadkhas fulfilled with intensity of urge to fully know the ancient wisdom of 'Arithmetic to astronomy range', and further about the phenomenon of transcendental carriers embedded with Brahman intelligence, are impressed upon to have proper initiation for conceptual foundation of basic operations the Vedic mathematics way of having sequential formats for numerals being of single double, triple and multiple digits.

For further intellectual chase

1. Intellectual history of man deserve to be written with a focus upon the source of concepts with their moments of emergence along the time line.
2. Credit goes to the Vedic system for the basic conceptual comprehension formats of 'zero, place value system and numerals with standard forms, frames and formats'.
3. With these three basic concepts (i) zero (ii) one (iii) numerals for the place value system standing traced to Vedic India, the chase further is to be for the geometric dimensional frames.

■

2

EXTENDED RANGE OF NUMERALS

1. Vedic system have extended range of numerals as ekadhiken numerals, eknunena numerals, dashadhika numerals and negative numerals.
2. This way the numerals range stands extended five folds and the availability of 49 numerals (with positive and negative zero numeral being of identical value), the same have been made of order of the coordination arrangements of axes format for the coverage of all the seven states of consciousness for their spatial order within Creator's space.
3. The ekadhikena numeral may be defined as a numeral of one count more. Illustrative the zero count would, as ekadhika numeral would become of value 0+1=1. likewise the numerals 0, 1, 2, 3, 4, 5, 6, 7, 8, 9 would become of values 1, 2, 3, 4, 5, 6 , 7, 8, 9 and 10 respectively. These ekadhika numerals are allotted special symbols as concerned numerals with dot at the top, as is illustrated depicted here:

Ekadhiken numerals									
•	•	•	•	•	•	•	•	•	•
1	2	3	4	5	6	7	8	9	0

4. Likewise the ekanunena numerals are defined as of values of one count less than that of the value of the concerned numerals. The specific symbols allotted are the symbols of the concerned numerals with dot at the bottom thereof as follows.

Ekanunena numerals									
1	2	3	4	5	6	7	8	9	0
•	•	•	•	•	•	•	•	•	•

5. Further the range of numerals is extended by the vedic system by inventing dashadhika numerals which by definition are of values of ten counts more than that of the original value of the numerals. The special symbols accepted by dashadhika numerals with dot on their left side, as to indicate as that it is to carry the value of the numeral with addition of ten counts as to be availing the next place value. The symbolic expressions table is being reproduced here under for these dashadhika numerals.

Dashadhika numerals									
•1	•2	•3	•4	•5	•6	•7	•8	•9	•0

6. The Vedic systems have further added to the richeness of their operation by inventing the range

of numerals with addition to it the range of negative numerals, which by definition accept one count less than that of the original values of the numerals. The special symbol accepted by them is of having a symbol of – at the top of the symbol are concerned of the numerals. Here under is being reproduced tabulation of symbolic expression of negative numerals.

Negative numerals								
$\bar{1}$	$\bar{2}$	$\bar{3}$	$\bar{4}$	$\bar{5}$	$\bar{6}$	$\bar{7}$	$\bar{8}$	$\bar{9}$

7. As such the extended range of numerals is being tabulated for facility of comprehensive view and for the insight into the processing range in terms of this extended range of numerals:

1	Numerals	1	2	3	4	5	6	7	8	9	0
2	Ekadhiken numerals	$\dot{1}$	$\dot{2}$	$\dot{3}$	$\dot{4}$	$\dot{5}$	$\dot{6}$	$\dot{7}$	$\dot{8}$	$\dot{9}$	$\dot{0}$
3	Ekanunena numerals	$\underset{\bullet}{1}$	$\underset{\bullet}{2}$	$\underset{\bullet}{3}$	$\underset{\bullet}{4}$	$\underset{\bullet}{5}$	$\underset{\bullet}{6}$	$\underset{\bullet}{7}$	$\underset{\bullet}{8}$	$\underset{\bullet}{9}$	$\underset{\bullet}{0}$
4	Negative numerals	$\bar{1}$	$\bar{2}$	$\bar{3}$	$\bar{4}$	$\bar{5}$	$\bar{6}$	$\bar{7}$	$\bar{8}$	$\bar{9}$	
5	Dashadhika numerals	•1	•2	•3	•4	•5	•6	•7	•8	•9	•0

8. Here, an illustrative case of applications of positive and negative numerals achieving the placement of bigger numerals, six, seven, eight and nine, into smaller absolute value numerals of the range of 0

to 5 and –1 to –4. This skill is known as 'vinculum'. This conversion of vinculum technique, is of specific values for writing tables.

■

3
VEDIC MATHEMATICS THE VEDIC WAY

3.1 INTRODUCTORY

1. 1 Vedic mental calculator, as human mind, deserves to be approached the Vedic way.

1.2.1 Vedic way to approach as a system is to follow parallely the two fold established paths of ***Sankhiya nishtha*** and ***Yoga nishta***.

1.3 This, as a processing process, would mean to permit the mind to transcend the manifestation format by availing the artifices (of numbers parallel to the dimensional frames).

1.4 The par-excellence feature of this system is that it has its beginning, as well as the end, and also the whole range in between, at the same 'transcendental ment', which makes it a self referral system unfolding from within by availing the Creator's space (4-space).

1.5 This transcendental ment is the 'Om iti Ek Akshar Braham' / The transcendental 'Om' being the sole syllable Brahman.

1.6 The sankhiya nishtha processing along the

'transcendental meant' unfolds quarter by quarter and each quarter itself to be of the order of the 'whole itself', with whole as first quarter being 'Om itself', and as second quarter being 'Om as Pranava', and steps ahead, as third quarter being Om as Aum and as fourth quarter, 'Om being Omkar'.

1.7 The yoga nishta processing beginning with surya namaskar / salute to the sun, and, initiations with the nad / primordial sound of 'Om' permits the mind to continue transcending quarter by quarter and transcending ahead of the fourth quarter as along the transcendence range of quarters of quarters as 'Om as Udgith' and a step ahead along quarters of quarters of quarters as 'Om as Vashtkar' and thereby attaining the skyline of transcendental worlds and glimpsing the inner folds of the transcendental worlds as well as the self referral core of the transcendental worlds.

1.8 This parallel processing along the two fold established paths of Sankhiya nishtha and yoga nishtha shall be perfecting intelligence of the sadkhas with which they to attain a take off stage for processing in terms of the ordering principle of Ganita Sutra-1 'Ekadhikena Purvena' / 'One more than before', which shall be ordering the 'affine state' into sequential order state 'with which the phase and stage of take off from the artifice format of 'one' would take to the artifice format of 'two' as one more than the one before, which would manifest as the functional rule of Ganita Upsutra-1 Anurupeyena / to follow the proportionate symmetry of forms within frames.

1.9 The sequential chase of the functional rules of Ganita Sutras (and Upsutras) shall be equipping the field of intelligence of the transcending mind availing artifices of numbers parallel to the dimensional frames with the potentialities of the order of 'VEDIC MENTAL CALCULATOR'.

1.10The sadkhas fulfilled with an intensity of urge to equip their intelligence fields with potentialities of 'VEDIC MENTAL CALCULATOR' shall be through the functional formats of Ganita Sutras (and Upsutras) in the sequence and order of the their text.

3.2 SEQUENTIAL INITIAL LEARNING STEPS

The initial learning steps, may be enumerated as under:

(1) Comprehension of count 'one'.

(2) Counting.

(3) Reverse counting

(4) Shift to ten place value system for counts as numbers

(5) Numerals

(1, 2, 3, 4, 5, 6, 7, 8, 9)

(6) Place value numeral (0)

(7) Placements values of numerals

10^0, 10^1, 10^2, 10^3, and so on.

(8) $10^0=1$

(9) Double digit expressions for numerals

(00, 01, 02, 03, 04, 05, 06, 07, 08, 09)

1	Numerals	1	2	3	4	5	6	7	8	9	0
2	Ekadhiken numerals	$\dot{1}$	$\dot{2}$	$\dot{3}$	$\dot{4}$	$\dot{5}$	$\dot{6}$	$\dot{7}$	$\dot{8}$	$\dot{9}$	$\dot{0}$
3	Ekanunena numerals	$\underset{\bullet}{1}$	$\underset{\bullet}{2}$	$\underset{\bullet}{3}$	$\underset{\bullet}{4}$	$\underset{\bullet}{5}$	$\underset{\bullet}{6}$	$\underset{\bullet}{7}$	$\underset{\bullet}{8}$	$\underset{\bullet}{9}$	$\underset{\bullet}{0}$
4	Negative numerals	$\bar{1}$	$\bar{2}$	$\bar{3}$	$\bar{4}$	$\bar{5}$	$\bar{6}$	$\bar{7}$	$\bar{8}$	$\bar{9}$	
5	Dashadhika numerals	•1	•2	•3	•4	•5	•6	•7	•8	•9	•0

Conversion of bigger numerals (6, 7, 8, 9) into smaller absolute values in terms of positive numerals 1 to 5 and negative numerals (-1, -2, -3 and –4) by Vinclum system

(1) Addition operation of numerals

(2) Multiplication operation of numerals

(3) Addition and subtraction operations

(4) Multiplication operation of two digits numbers

(5) Tables 99 x 99

(6) Multiplication operation of three digits and larger digits numbers (which better may be left for next phase and stage of learning)

(7) Division by two digits divisors

(8) Sorting of primes amongst two digits numbers

(9) Division by three digits and larger digits numbers (which better may be left for next phase and stage of learning)

For further intellectual chase

1. The credit for the comprehension of pair of opposite

orientations of 'interval', and as being of 'distinct' dimensional formats also goes to the Vedic systems, as much as that the pair of orientations pair parallel to the pairing of (+1) space and (–1) space.

2. This comprehension of Vedic systems goes to the route of dimensional set ups, as that (-1 space) plays the role of dimension of (+1) space.
3. And, in general, (N-2) space playing the role of dimension of N-space.

3.3 NEXT PHASE AND STAGE OF LEARNING

The initial learning with focus upon the ordering principle for coordination arrangement of artifices of double digits numbers along ten place value system in respect of the basic operations, shall be creating a phase and stage for further learning with focus upon the geometric formats.

The sequential learning steps for this phase and stage of learning, as a middle schooling of Discipline of mathematics, may be as under:

(1) To have a conscious note as that the focus of the learning of this phase and stage is going to be centered around the set up of a cube.

(2) To learn as that 'cube' exists in continuity of 'interval' and square.

(3) To comprehend the set up of cube as enveloped within a synthetic boundary stitched in terms of eight corner points, 12 edges and 6 surfaces.

(4) To chase the synthetic set up of the boundary of the cube, firstly as a flow stream connecting all the 8 corner points in seven steps, and that there

are as many as 120 such streams running through the edges and corners of the cube, secondly, as that the surfaces accept grid frames for flow of linear order within the domain of the cube and thirdly as that the center of the domain is equidistant form the corners and is the intersection of four planes, and that each such plane is of distinct dimensional existence than those of 3 planes containing the three linear dimension of the three dimensional frame, and finally that, though cube is the representative regular body of 3-space but at the same time it accepts manifestation format of Creator's space (4-space).

(5) Of the above four different aspects of learning, the focus during present phase and stage of learning is to concentrate more upon the first two aspects only, and the remaining two aspects, are just to be introduced for initial exposure only, and the in depth learning about these two aspects, namely third and fourth aspects mentioned above, is to be the subject matter of learning at high school level.

(6) The attainment of learning of this phase and stage is to be of the order of smooth transition from the set up of a cube to the set up of hyper cube-4.

(7) For ensuring smooth transition for learning from the present phase and stage of middle school level mathematics focusing upon the set up of a cube as representative regular body of 3-space to next phase and stage of learning with focus upon hyper cube 4 as representative regular body of 4-space,

one is to shift for its focus from the set up of a cube to a set up of a sphere.

(8) The basic difference of the set up of the sphere from that of the set up of the cube is that while in case of the cube it has synthetic boundary of as many as six surface plates and the volume is accepted as a single unified entity. On the other hand, in case of a sphere, the boundary is a single unified surface while within it the domain is of fragmented set up which is evident from the values for its volume being of units, area of the surface x 1 / 3.

3.4 VEDIC MENTAL CALCULATOR

For initial stage learners, designated as young minds, the Vedic mental calculators operational steps are restricted to numbers of ten place value system, and of these as well, are being intentionally restricted up till two digits (numbers).

General format of multiplication at first place value of the place value system	0 1 x 0 2
First number LM: L M Second number AB: A B L x A | (L x B) + (M x A) | M x B (L x A) + ((L x B) + (M x A)) + (M x B)	First number 0 1 0 1 =0+1 Second number 1 1 0 2= 1+1 0 x 1 | (1 x 1) + (0 x1) | 1 x 1 0 | 1 | 1 0 + 1 + 1 =2

Further, out of four basic operations of addition, subtraction, multiplication and division, only 'multiplication and division' are being taken up, and the addition and subtractions, being otherwise as very simple, and moreover these being just direct and reverse counting, as such these are to be dealt with as extension of counting.

MULTIPLICATION

Multiplication of numerals of double digits formats

1. It be accepted as definitions as that

   1 × 1 =1, 0 × 1 =0, 1 × 0 = 0, 0 × 0=0.

2. The numerals, 1, 2, 3, 4, 5, 6, 7, 8, 9, 0, are to have placements in first place of place value system, and as such the addition is to follow the rule at each step 'of one more'. And 1+1=2, 1+2=3,– 1+8=9, to depict the internal structure of the numerals with placements at first place of the place value system.

3. The Vedic mental calculator for chase of multiplication of numerals on first principles, follows the procedure of skills as steps as depicted and detailed here under:

   (i) Presumes the intial results '1 × 1 =1, 0 × 1 =0, 1 ×0=0, 0 × 0=0'.

   (ii) Adopts a skill of first converting the numerals into double digits format as 01, 02, 03, 04, 05, 06, 07, 08, 09,

   (iii) And then further re-expressing them as:

   01=0+1, 02=0+2, 03=0+3, 04=0+4, 05=0+5, 06=0+6, 07=0+7, 08=0+8, 09=0+9, 00 = 0+0

   (iv) Then, the format of cross is availed for setting of the digits of the numbers LM and AB (here

01 and 02) to be multiplied, as formatted above.

(v) This being the general format for multiplications at the first place of the place value system, hence sequentially the values of products of numerals amongst themselves can be obtained as in case of 01 × 02 = LM × AB.

MENTAL PROCEDURE

2. For the above procedural steps and the final value of the product 01 × 02 = LM × AB, to be approached 'as oral / mental computations', the procedure may be sequenced as under

    (1) 'The inner surface of the 'FOREHEAD' be taken as the mental board.

    (2) The above format of cross with settings for the digits be had for the reference facility of the brain.

    (3) As memory facility indicator for the brain, the joint of the cross be taken as it is along the nose line through its root at the mid of the eyes line and extended ahead through the forehead board, is to be taken as a line of division between the digits as left eye digit and right eye digit.

    (4) Once the format of cross and setting of the digits is made lively as above on the forehead inner surface board, the further steps are to be to have,

Firstly the multiplication of right eye digits namely, M & B, and product value be obtained as M x B.

Then the pair of cross multiplication of upper right eye digit with lower left eye digit and of upper left eye digit with lower right eye digit be had as L x B and M x A.

Finally the multiplication of left eye digits to be obtained as LA.

Once these four multiplications / products values are obtained then the only step remain to be done is to have their summation as (M x B) + {(L x B) + (M x A)} + (L x A).

3. The above narration and description of the procedural details for setting and steps along a format at forehead inner surface board, though at first reading may look like a complex set up, but in fact it is so simple that once one is through it, it would become blissfully lively for the mind to transcend through it and to help the brain to carry out the steps without any strain what so ever.

4. It would be a blissful exercise for the young minds to use their mental calculator as of the order of Vedic mental calculator and to have perfection of intelligence for handling the multiplications of numerals.

MULTIPLICATION OF TWO DIGITS NUMBERS

1. Once one has perfected ones intelligence for using Vedic mental calculator for multiplication of numerals, the further application of this calculator for multiplication of double digit numbers would be nothing but repetitions of the above steps with the only difference that the three values 'namely (L x A), {(L x B) + (M x A)}, and (M x B) are not to be

added, and instead these are to be taken as values at three sequentially arranged places of the place value system.

2. It may be chased for the pair of digits 11 and 12. the mental setting on the inner surface of the forehead would give status as

   L=1(upper left eye digit), M=1 (upper right eye digit), A=1 (lower left eye digit), B=2 (lower right eye digit)

3. The left side product value would be L x A = 1 x 1=1, the middle pair of products values (1 x 2) and (1 x 1) and the right side product value would be (1 x 2).

4. These three sets of values, namely '1', '2 +1', and '1', when to have the setting of places of ten place value system these shall be yielding the product value for 11 x 12 as 132.

5. Following the above procedure one can workout whole range of product of any pairs of two digits numbers. However one additional care which is to be taken when the product values of individual digits is of value more than 10, then there is to be a carry forward step which need be taken. Illustratively when 27 and 29 are to be multiplied, the right eye digits, namely 7 and 9 shall be yielding product value as 63, and it being a value more than 10. as such '6' is to be carried forward to the place ahead.

Complete mental steps would be to obtain

(i) Product of the left eye digits namely 2 x 2 =4

(ii) Sum of the cross products, namely 2 x 9+ 2 x 7=32.

(iii) Product of right eye digits, namely 7 x 9=63.

The placements for these three values would be as follows

10^3	10^2	10^1	10^0
Fourth Place	Third Place	Second Place	First Place
	4	32	63

Now we can see that the first place is having value 63, as such it would provide '6' to be carried forward. Once it is carried forward the places values would be as follows

10^3	10^2	10^1	10^0
Fourth Place	Third Place	Second Place	First Place
	4	38	3

Now with this the middle place value is having entry as '38' and as such it provide '3 to be carried forward. Once it is carried forward the places values would be as follows.

10^3	10^2	10^1	10^0
Fourth Place	Third Place	Second Place	First Place
	7	8	3

As such 27 x 29= 783.

The whole process amounts to product of single digits, and the second digit of place values to be carried forward. Both these steps are simple enough to be easily performed orally as mental steps and as such one can develop up to date the skills and attain mental calculator of the order of Vedic mental calculator for handling multiplications of double digits numbers.

This skill can be very gently extended to three digits numbers, four digits numbers and even to large number of digits but for the present phase and stage of learning, the perfection of intelligence for handling pair of digits would be laying down firm foundations for further learning and skills needed.

10^3	10^2	10^1	10^0
Fourth Place	Third Place	Second Place	First Place
	4	32	63

TABLES

1. Counting is just a table of '1'.
2. The table of '1' (table '1') is known as counting table.

1.	2.	3.	4.	5.	6.	7.	8.	9.	10.
11.	12.	13.	14.	15.	16.	17.	18.	19.	20.
21.	22.	23.	24.	25.	26.	27.	28.	29.	30.
31.	32.	33.	34.	35.	36.	37.	38.	39.	40.
41.	42.	43.	44.	45.	46.	47.	48.	49.	50.
51.	52.	53.	54.	55.	56.	57.	58.	59.	60.
61.	62.	63.	64.	65.	66.	67.	68.	69.	70.
71.	72.	73.	74.	75.	76.	77.	78.	79.	80.
81.	82.	83.	84.	85.	86.	87.	88.	89.	90.
91.	92.	93.	94.	95.	96.	97.	98.	99.	100.

3. As here we are interested to learn to write tables '1 to 100' of 100 steps each, as such, let us, first of all tabulate here under the counting up till 100, and to designate it as table '1', that is, 1 x 100.

4. Vedic systems avail positive and negative numerals simultaneously to simplify the operations in terms of smaller value numerals, namely, 1, 2, 3, 4 & 5. The bigger numerals namely 6, 7, 8 & 9 are replaced as

$$6=1\bar{4},\ 7=1\bar{3},\ 8=1\bar{2},\ 9=1\bar{1}$$

5. Availing this technique, the above counting table is rewritten as under

1	2	3	4	5	$1\bar{4}$	$1\bar{3}$	$1\bar{2}$	$1\bar{1}$	10
11	12	13	14	15	$2\bar{4}$	$2\bar{3}$	$2\bar{2}$	$2\bar{1}$	20
21	22	23	24	25	$3\bar{4}$	$3\bar{3}$	$3\bar{2}$	$3\bar{1}$	30
31	32	33	34	35	$4\bar{4}$	$4\bar{3}$	$4\bar{2}$	$4\bar{1}$	40
41	42	43	44	45	$5\bar{4}$	$5\bar{3}$	$5\bar{2}$	$5\bar{1}$	50
51	52	53	54	55	$1\bar{4}\bar{4}$	$1\bar{4}\bar{3}$	$1\bar{4}\bar{2}$	$1\bar{4}\bar{1}$	$1\bar{4}0$
$1\bar{4}1$	$1\bar{4}2$	$1\bar{4}3$	$1\bar{4}4$	$1\bar{4}5$	$1\bar{3}\bar{4}$	$1\bar{3}\bar{3}$	$1\bar{3}\bar{2}$	$1\bar{3}\bar{1}$	$1\bar{3}0$
$1\bar{3}1$	$1\bar{3}2$	$1\bar{3}3$	$1\bar{3}4$	$1\bar{3}5$	$1\bar{2}\bar{4}$	$1\bar{2}\bar{3}$	$1\bar{2}\bar{2}$	$1\bar{2}\bar{1}$	$1\bar{2}0$
$1\bar{2}1$	$1\bar{2}2$	$1\bar{2}3$	$1\bar{2}4$	$1\bar{2}5$	$1\bar{1}\bar{4}$	$1\bar{1}\bar{3}$	$1\bar{1}\bar{2}$	$1\bar{1}\bar{1}$	$1\bar{1}0$
$1\bar{1}1$	$1\bar{1}2$	$1\bar{1}3$	$1\bar{1}4$	$1\bar{1}5$	$10\bar{4}$	$10\bar{3}$	$10\bar{2}$	$10\bar{1}$	100

6. We can note that the above counting table avails the following numerals

Zero	0
positive numerals	1, 2, 3, 4, 5
negative numerals	$\bar{1}, \bar{2}, \bar{3}, \bar{4}$

7. There is table (1) corresponding to table (1); the table 1 is counting table, while table (1) is reverse counting. It also may be designated as reverse counting table or simply as negative counting.
7. Likewise the tables of (2, 3 & 4) may be designated as negative tables corresponding to tables (2, 3, 4) respectively.
8. Therefore the essential pre-requisite knowledge for writing any table up till any number of steps is the values of tables 1, 2, 3, 4 & 5 only. The negative tables would follow of their own as negative values.
9. The tables 1 to 100 can be written with the help of the knowledge of tables 1, 2, 3, 4 & 5 as every number from 1 to 100, as is evident from the above counting table no-2, can be written as 1, 2 or 3 digits form, and each digit to flow out corresponding value table.
10. This way, the one digits number namely 1, 2, 3, 4, 5 shall be flowing out tables 1, 2, 3, 4, 5 and the same can be straight a way be written with the help of the previous knowledge.
11. The table for number 14, being a double digit expression, can be written for values in two parts, first as values of table 1 and second as values of table 4, that is negative values of table 4.
12. The flow chart for this table would follow as under:

Step	First Part	Second Part	Value of Step
Step-1:	1×6	1 $\bar{4}$	10-4=6
Step-2	2×6	2 $\bar{8}$	20-8=12
Step-3	3×6	3 $\bar{1}\bar{2}$	30-12=18
Step-4	4×6	4 $\bar{1}\bar{6}$	40-16=24

13. The above steps values are worked out, firstly computing the values for the first part as table-1 and for the second part as table 4 as negative values of table 4, and then the simplification. Here, for mental operations, the knowledge of table 4 is presumed.
14. Likewise the table 13 can be written as under

Step	First Part	—Second Part	Value of Step
Step-1:	1×7	1 $\bar{3}$	10-3=7
Step-2	2×7	2 $\bar{6}$	20-6=14
Step-3	3×7	3 $\bar{9}$	30-9=21
Step-4	4×7	4 $\bar{1}\bar{2}$	40-12=28

15. Likewise the tables $\overline{12}$ & $\overline{11}$ can be written as under:

Table $1\overline{2}$

Step	First Part	Second Part		Value of Step
Step-1:	1×8	1	$\overline{2}$	10-2=8
Step-2	2×8	2	$\overline{4}$	20-4=16
Step-3	3×8	3	$\overline{6}$	30-6=24
Step-4	4×8	4	$\overline{8}$	40-8=32

Table $\overline{11}$

Step	First Part	Second Part		Value of Step
Step-1:	1×9	1	$\overline{1}$	10-1=9
Step-2	2×9	2	$\overline{2}$	20-2=18
Step-3	3×9	3	$\overline{3}$	30-3=27
Step-4	4×9	4	$\overline{1}$ 4	40-4=36

16. Now if we have to write tables for 96, 97, 98 and 99, then first of all we shall be rewriting them as $10\overline{4}$, $10\overline{3}$, $10\overline{2}$ and $10\overline{1}$. Then, as these are 3 digits

expressions, so each digit shall be making its independent contribution at each step of the table. Accordingly the values of table are to received in three parts.

Illustratively 10$\bar{4}$ shall be flowing out, first part as table 1 values, second part as table 0 values and third part as negative of table 4 values. Therefore our knowledge of tables 0, 1 and 4 shall be helping us to flow out the table of 96 as under

Table 96= Table 10$\bar{4}$

Step	First Part	Second Part	Third Part	Value of Step
1x96	1	0	$\bar{4}$	100-0-4=96
2x96	2	0	$\bar{8}$	200-0-8=192
3x96	3	0	$\bar{1}\bar{2}$	300-0-12=288
4x96	4	0	$\bar{1}\bar{6}$	400-0-16=384

Table 97= Table 103

Step	First Part	Second Part	Third Part	Value of Step
1x97	1	0	$\bar{3}$	100-0-3=97
2x97	2	0	$\bar{6}$	200-0-6=194
3x97	3	0	$\bar{9}$	300-0-9=291
4x97	4	0	$\bar{1}\bar{2}$	400-0-12=388

Table 98= Table $10\bar{2}$

Step	First Part	Second Part	Third Part	Value of Step
1x98	1	0	$\bar{2}$	100-0-2=98
2x98	2	0	$\bar{4}$	200-0-4=196
3x98	3	0	$\bar{6}$	300-0-6=294
4x98	4	0	$\bar{8}$	400-0-8=392

Table 99= Table $10\bar{1}$

Step	First Part	Second Part	Third Part	Value of Step
1x99	1	0	$\bar{1}$	100-0-1=99
2x99	2	0	$\bar{2}$	200-0-2=198
3x99	3	0	$\bar{3}$	300-0-3=297
4x99	4	0	$\bar{4}$	400-0-4=396

Tables of numbers of digits 0, 1, 1

2.1 The following table converts number 1 to 100 into smaller digits expressions.

1	2	3	4	5	$1\bar{4}$	$1\bar{3}$	$1\bar{2}$	$1\bar{1}$	10
11	12	13	14	15	$2\bar{4}$	$2\bar{3}$	$2\bar{2}$	$2\bar{1}$	20
21	22	23	24	25	$3\bar{4}$	$3\bar{3}$	$3\bar{2}$	$3\bar{1}$	30
31	32	33	34	35	$4\bar{4}$	$4\bar{3}$	$4\bar{2}$	$4\bar{1}$	40
41	42	43	44	45	$5\bar{4}$	$5\bar{3}$	$5\bar{2}$	$5\bar{1}$	50
51	52	53	54	55	$1\bar{4}\bar{4}$	$1\bar{4}\bar{3}$	$1\bar{4}\bar{2}$	$1\bar{4}\bar{1}$	$1\bar{4}0$
$1\bar{4}1$	$1\bar{4}2$	$1\bar{4}3$	$1\bar{4}4$	$1\bar{4}5$	$1\bar{3}\bar{4}$	$1\bar{3}\bar{3}$	$1\bar{3}\bar{2}$	$1\bar{3}\bar{1}$	$1\bar{3}0$
$1\bar{3}1$	$1\bar{3}2$	$1\bar{3}3$	$1\bar{3}4$	$1\bar{3}5$	$1\bar{2}\bar{4}$	$1\bar{2}\bar{3}$	$1\bar{2}\bar{2}$	$1\bar{2}\bar{1}$	$1\bar{2}0$
$1\bar{2}1$	$1\bar{2}2$	$1\bar{2}3$	$1\bar{2}4$	$1\bar{2}5$	$1\bar{1}\bar{4}$	$1\bar{1}\bar{3}$	$1\bar{1}\bar{2}$	$1\bar{1}\bar{1}$	$1\bar{1}0$
$1\bar{1}1$	$1\bar{1}2$	$1\bar{1}3$	$1\bar{1}4$	$1\bar{1}5$	$10\bar{4}$	$10\bar{3}$	$10\bar{2}$	$10\bar{1}$	100

2.2 The numbers availing digits 0, 1, 1 are being remarked with red colour for pointed attention

2.3 The table of these numbers would follow by just addition or subtraction of values 0 or 1 at each step.

2.4 Illustratively of 111=89 can be written just by addition of 1 at each step of column no 3 and subtraction of 1 at each step of columns no 1 and 2. The entries would be following

1	2	3	4	5	$1\bar{4}$	$1\bar{3}$	$1\bar{2}$	$1\bar{1}$	10
11	12	13	14	15	$2\bar{4}$	$2\bar{3}$	$2\bar{2}$	$2\bar{1}$	20
21	22	23	24	25	$3\bar{4}$	$3\bar{3}$	$3\bar{2}$	$3\bar{1}$	30
31	32	33	34	35	$4\bar{4}$	$4\bar{3}$	$4\bar{2}$	$4\bar{1}$	40
41	42	43	44	45	$5\bar{4}$	$5\bar{3}$	$5\bar{2}$	$5\bar{1}$	50
51	52	53	54	55	$1\bar{4}\bar{4}$	$1\bar{4}\bar{3}$	$1\bar{4}\bar{2}$	$1\bar{4}\bar{1}$	$1\bar{4}0$
$1\bar{4}1$	$1\bar{4}2$	$1\bar{4}3$	$1\bar{4}4$	$1\bar{4}5$	$1\bar{3}\bar{4}$	$1\bar{3}\bar{3}$	$1\bar{3}\bar{2}$	$1\bar{3}\bar{1}$	$1\bar{3}0$
$1\bar{3}1$	$1\bar{3}2$	$1\bar{3}3$	$1\bar{3}4$	$1\bar{3}5$	$1\bar{2}\bar{4}$	$1\bar{2}\bar{3}$	$1\bar{2}\bar{2}$	$1\bar{2}\bar{1}$	$1\bar{2}0$
$1\bar{2}1$	$1\bar{2}2$	$1\bar{2}3$	$1\bar{2}4$	$1\bar{2}5$	$1\bar{1}\bar{4}$	$1\bar{1}\bar{3}$	$1\bar{1}\bar{2}$	$1\bar{1}\bar{1}$	$1\bar{1}0$
$1\bar{1}1$	$1\bar{1}2$	$1\bar{1}3$	$1\bar{1}4$	$1\bar{1}5$	$10\bar{4}$	$10\bar{3}$	$10\bar{2}$	$10\bar{1}$	100

The pattern can be availed for all the above three forms for each step. It may be noticed that in each column there is a sequential increase or decrease of value '1'. While in column 1 and 2 there is a sequential decrease of value 1 and in column 3 there is a sequential increase of value 1. As such the table 89 can be straight a way mentally written just by following the increase, decrease rule from the re expression form of 89 as $1\bar{1}\bar{1}$ indicating that column 1 and 2 would follow decrease of value 1 while column 3 would follow increase of value 1,

Note: The other cases of numbers availing these three digits namely 0, 1 and 1 may be worked out for their tables up till any step as exercises.

Note: The senior students can workout the patterns for the other numbers as well. As such the formal write ups of lessons 3 to 7 are being held up for the present. So that students may have an opportunity to work these out of their own. However if any student still wants help may share it with us

-- DIVISION

The Vedic system approaches the division operation of any number by any other number in most natural digit by digit approach.

The division by single digit number, which means by numerals (1 to 9), is just a counting game or a skill of tables of this range, and as such, it be better mentally approached as such.

The division by divisors of two digits as a straight mental calculations, need be learnt the Vedic mental calculator way. Here below, the 'highest ideal' of the Vedic Sutras, as brought to focus by Swami Bharti Krsna Tirtha Ji Maharaj, is being illustrated for glimpsing its bliss and for young minds to perfect their skills accordingly.

Swami Ji demonstrated the wonders of the Ganita Sutras of straight division, taking the illustrative case of 38982/ 73.

The entire division is done only in terms of the numeral '7', and it is this what makes all the difference of the Vedic system over the present day system which would be requiring division by '73'. The numeral '7' can be very conviently handled for its multiples while the multiples of '73' would require much stress for the mind to reach at.

Swami ji has given the details of the steps of straight division of the illustrative case with algebraic proof thereof.

The dividend 39982 is re expressed in algebraic equation format as $38x^3+9x^2+8x+2$; and the divisor 73 is expressed as $7x+3$. Then the divisions steps of $38x^3+9x^2+8x+2$ by $7x+3$ are chased, and the same follow as:

$7x+3)\ 38x^3+9x^2+8x+2(5x^2+3x+4$	
$35x^3+15x^2$	--Step-1, to mutiply divisor by $5x^2$
$3x^3-6x^2$	--Step-2, to subtract value of step-1 and to carry forward 3x from first item to the second item as $30x^2$
$= 24x^2+8x$	--Step-3 and bring down 8x from the dividend
$\underline{21x^2+9x}$	--Step-4 to mutiply divisor by 3x
$3x^2-x$	--Step-5 to subtract value of step-4, and to carry forward 3x from first item to the second item as 30x
$= 29x+2$	--Step-6 and bring down +2 from the divident
$\underline{28x+12}$	-- Step 7 to multiply the divisor by 4
$x-10$	--Step 8 to subtract value of step 7 and to carry forward x from item no 1 to the second item as 10
$= \underline{10-10}$	--Step-9 and to reach at 10-10=0 and with it the process is complete as all the terms of the divident have been completly downloaded and taken through the process of division
$= \underline{0}$	

This as such yields quotient being '$5x^2+3x+4$', which on conversion to arithmetic format of ten place value system comes to be 534 and the reminder being '0'.

This, this way, makes the process of reaching 'at quotient' and remainder, straight a way, as a mental steps of rolling down digit by digit.

For it, the first the divisor (73) is split into two parts, the first as, divisor digit (7) and second as flag digit (3).

Then the divident (38992) is split into two parts, the main quotient portion (3898) and the remainder portion (2). The remainder portion is to be of digits equal to the digits of (divisor digit (s)), which here is a single digit (7), and as such the remainder portion is a single digit portion (2)

As such, the divident portions and divisor digits are placed in the following format for further division steps:

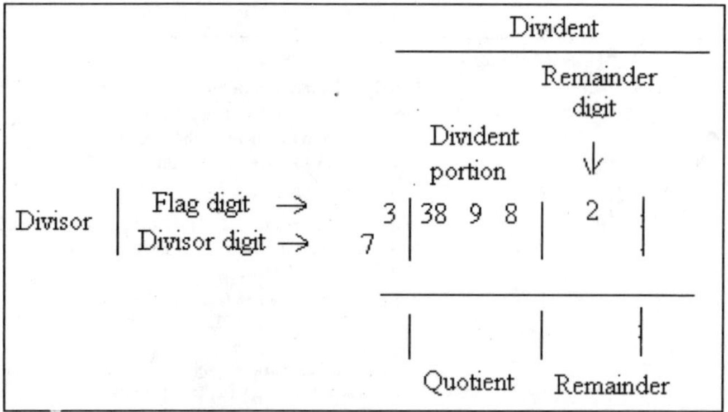

Here is being expressed, in the language of Swami ji as to the details of actual mental division steps for this illustrative case at hand:

"We divide 38 by 7 and get 5, as the quotient and 3 as the remainder. We put 5 down as the first quotient-digit and just prefix the remainder 3 up before the 9 of the dividend. In other words, our actual second-step gross dividend is 39. From this, we however, deduct the product of the indexed 3 and the first quotient digit 5, i.e. 3 x 5 = 15. The remainder 24 is our actual net-dividend. It is then dividend by 7 and gives us 3 as the second quotient-digit and 3 as the remainder; to be placed in their respective places as was done in the first step. From 38 the gross dividend thus formed, we subtract 3x the second quotient-digit 3, i.e. 9, get the remainder 29 as our next actual dividend and divide

that by 7. We get 4 as the quotient and 1 as the remainder. This means our next gross dividend is 12 from which, as before, we deduct 3 x the third quotient-digit 4, i.e. 12 and obtain 0 as the remainder. Thus we say : Q is 534 and R is zero. And this finishes the whole procedure; and all of it is one-line mental Arithmetic in which all the actual division is done by the simple-digit Divisor 7."

For facility of comprehension, the above procedure steps are being phased here under, which being followed in the light of the following placement chart.

			Divident	
			Divident portion	Remainder digit ↓
Divisor	Flag digit →	3	38 9 8	2
	Divisor digit →	7	3 3	1
			5 3 4	0
			Quotient	Remainder

PHASED PROCEDURE STEPS

(1) *We divide 38 by 7 and get 5, as the quotient and 3 as the remainder.*

(2) *We put 5 down as the first quotient-digit and just prefix the remainder 3 up before the 9 of the dividend.*

(3) *In other words, our actual second-step gross dividend is 39. From this, we however, deduct the product of*

the indexed 3 and the first quotient digit 5, i.e. 3 x 5 = 15.

(4) *The remainder 24 is our actual net-dividend. It is then dividend by 7 and gives us 3 as the second quotient-digit and 3 as the remainder; to be placed in their respective places as was done in the first step.*

(5) *From 38 the gross dividend thus formed, we subtract 3x the second quotient-digit 3, i.e. 9, get the remainder 29 as our next actual dividend and divide that by 7.*

(6) *We get 4 as the quotient and 1 as the remainder. This means our next gross dividend is 12 from which, as before, we deduct 3 x the third quotient-digit 4, i.e. 12 and obtain 0 as the remainder.*

(7) *Thus we say : Q is 534 and R is zero. And this finishes the whole procedure; and all of it is one-line mental Arithmetic in which all the actual division is done by the simple-digit Divisor 7."*

As mental calculator chase

The setting of dividend and divisor be maid lively on the forehead inner surface board as follows:

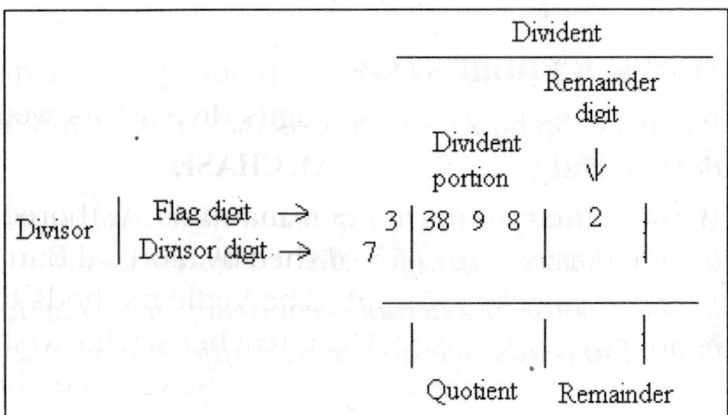

Then the first left hand digit of the quotient, namely, '5' be obtained by arranging 38 = 7 × 5 + 3. It remainder '3' be placed in front of the dividend digit '9'. The product of flag unit '3' and first quotient digit '5', namely 3 × 5 =15 be subtracted from 39and to reach at 39 + 15 = 24, as a divident for second step chase, and the procedure is to be continued mentally as 24 = 7 × 3 + 3. The second quotient digit would be obtained as '3'. The remainder digit '3' to be place in front of the divident digit '8'. The product value 3 x 3 to be subtracted from 38 and the divident for further division steps to be obtained as 38-9=29. Now for the third quotient digit, the above procedural step are to be repeated as 29 = 7 × 4+1. it would be yielding third quotient digit as '4' and the remainder digit '1' would get placement before quotient digit '2'. It would lead to 12-4 x 3=0. And the division is complete.

It may be evident that the whole division procedure is in fact nothing but division and multiplication of numerals and then simple addition and subtraction operations. A step by step chase in one or two illustrative cases would make ones mental calculator to be ready and that to of the order of Vedic mental calculator for all the divisors of two digits.

This range can be smoothly extended for three digits and even for four and higher digits divisors as well.

FOR FURTHER INTELLECTUAL CHASE

Three volumes of Teacher's manuals Authored by Kenneth R Williams (and published by Moti Lal Banarsi Dass, Delhi) may be referred. The Vedic methods have been brought within reach for teacher's who wish to teach the Vedic system. The message has been blissfully brought home as that Vedic mathematics is a complete system of mathematics, and that it has many surprising properties.

■

SECTION 2

FOUNDATIONS TO BE REVISITED

1

INTRODUCTORY

1. Beginning and end of 'skills' is appreciation and application of one different features of 'ONE'.
2. The proficiency of applied values of skills is straight a way linked with the comprehension of '**one**'.
3. Vedic systems accept 'one' as a transcendental set up and the same is of the order of the 'transcendental worlds'.
4. Ganita Sutras accept a transcendental prefix, 'Om' being 'Ek Akshar Braham' / Sole syallable Brahman.
5. The 'Ek / one' as designation and feature of 'Om', when availed, the same sets the 'Vedic mathematics system of Ganita Sutras' to unfold the coordination arrangements of organization formats for manifestation of creations as well as their transcendence (including ascendance) phenomenon availing 'artifices of numbers'.
6. As such, the students of mathematics, fulfilled with intensity of urge to acquire applications of skills of mathematics with proficiency, shall permit

their transcending mind to avail the artifices of Ganita Sutras for perfection of their intelligence of the order of the Ganita Sutras.

7. As such, first of all, a quick overview and glimpse of the sequence of steps of coordination arrangements of the functional formats of Ganita Sutras is being attempted so that in the light of it the intelligent applications of different Vedic mathematics skills may be leant.
8. The focus here during the overview and glimpse of the coordination arrangements of functional formats of the individual sutras together synthesizing a Vedic mathematics system has been upon the 'counts' as of 'points formats' and the counting numbers as sequential steps are the milestones along the track of a moving point.
9. The Ganita Sutras are being accepted as the first source reservoir of first principles for chase of 'basics' for acquiring proficiency of handling the Vedic mathematics skills in terms of artifices of numbers parallel to the geometric formats.
10. The functional rule of Ganita Sutra-1, namely 'one more than before', and the 'ordering principle' at its base with functional rule of 'pairing' become the basic of all basics.
11. The chandas (Vedanga) / science of meters is the source reservoir of pure and applied values of 'ordering principle' as well as of 'pairing operation' for structuring filters / meters / chandas.
12. The chandas (Vedanga) / science of meters works out the Vedic comprehension of 26 basic elements, permitting quarter by quarter chase and this as

such it takes to the basic 'tetra monad' approach of Vedic systems.

13. The another basic source of 'words formulations' availing 'paring rule' for 'ordering' organization formats with coordination of values along 'artifices of numbers parallel to geoemetric formats is the orthodox and classical English vocabulary.

14. As such the number value formats, is one such concept, which shall be proficiently learnt as a pre-requisite. In terms of this concept of number value formats, in short NVFs, each word (or even group of words) lead to the artifice of number availed by it. This artifice of number, as number value format, being availed by the 'word formulation' is nothing but the summation of the values of artifices of individual letters constituting the words. The individual letters of alphabet A to Z respectively availed artifice 1 to 26 in that sequence and order. The half of 26 as 13 and double of 26 as 52, being simultaneously available within Creator's space / 4-space as 2/1 x ½ = 1 x 1=1, as such the pairing operation being availed by the number value formats makes it of the order of processing quarter by quarter, and their by it also emerges to be running parallel to the 'quarter by quarter approach of chandas (Vedanga) / meters.

15. With it, (1) the proficiency of skill of counting and (2) the proficiency of pairing of counts, particularly counts 1 to 26, emerges to be the basic pre-requisite for proficiently handling the Vedic skills in terms of the Ganita Sutras principles.

16. In addition to the skill of proficient counting, which

may be designated as direct / increasing sequence of counting, the expectations of the skills from the students as a pre-requisite proficiency, is of reverse counting.

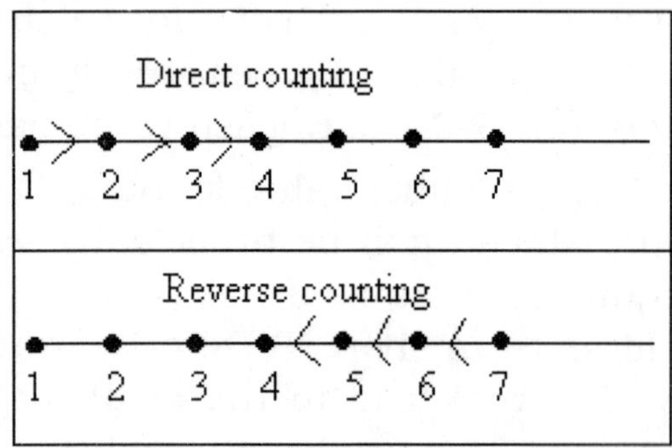

17. The direct and reverse counting, being a jumping exercise of one step at a time, but very is a necessity that the step should be of 'any particular value'. How it would be, if the jump step is of '2 counts', instead of 'one count' only.

18. The 'skills' starts unfolding like that, as the jumping steps of 'two counts' shall be taking us from 'counting' to a table of '2'.

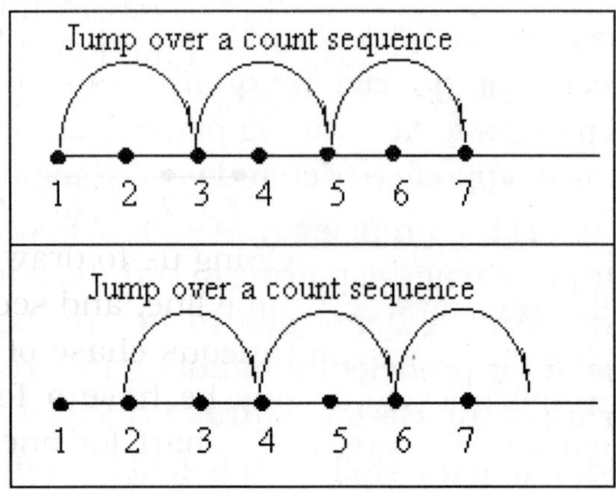

■

2
SEQUENTIAL SELF-EVALUATIONS

1. First skill to which one shall acquire is to reach at sequential self-evaluations of progress of learning.
2. For it, Ganita Sutra-1 provides us the 'ordering principles'. It is of a feature of a chase as 'one step more than that of the previous stage steps.
3. This rule, when followed would help us number and sequence the steps of learning attained. It would also make available the whole range of counting numbers and the numbers line format beneath for availing it as a 'time line' as well as the 'learning progress line'.

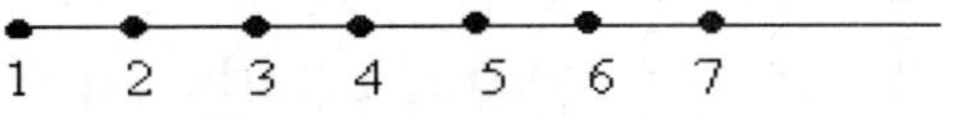

4. Further, this shall be helping us to draw a pair of parallel lines, first as a time line, and second as a progress line. By simultaneous chase of this pair of parallel lines, one may be have a format for 'sequential self evaluation' chart for oneself.

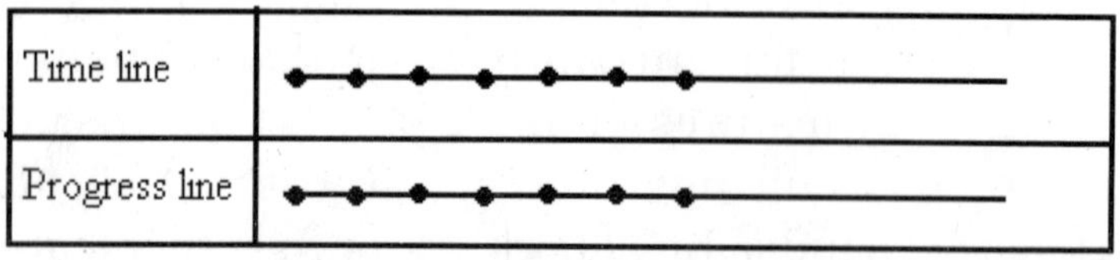

Time line	•—•—•—•—•—•—•————————
Progress line	•—•—•—•—•—•—•————————

5. The step by step, motion along this set of parallel lines, namely along time-line and progress-line, one may start having comprehension of one's sequential progress of learning.
6. The time-line shall be 'phasing the time' as 'time units', while the 'progress line' shall be requiring phasing of 'progress' as 'progress of learning' as 'learning units'.
7. The 'the learning units', as such, from the point of view of the Discipline of mathematics, as approached by 'Vedic mathematics system of Ganita Sutras' shall be focusing upon the very first 'learning unit' being 'one' itself.
8. Therefore this whole exercise of sequential self evaluation skill is to ultimately results into the 'comprehension of ONE', and it is to emerge, as is the enlightenment on the point as preserved in the available scriptures, to be as big and wide, as is the whole universe itself, of self referral features and a self sustaining set up, with its beginning as well as the end to be at the same 'ment'.
9. Accordingly the young minds shall educates themselves to enlarge their comprehension domain of the features of 'one'. For it, they may begin seeing as to what are the situation around which may be associated 'one' as a value or as an artifice, and the like. Illustratively 'one teacher', 'one book', 'a seat in the class room', etc. etc.

10. The other way, in which, young minds may initiate themselves for enriching their comprehension domain of features of 'one', may be, as that one may start enlisting different 'words', which may have been indicating about one or the other feature or features of 'ONE'. Illustratively, 'ONE', 'single', 'sole', 'whole', 'monad', and the like, as initial entries of the table of such word- formulations of meanings indicating one or other feature(s) of 'ONE' may be extending and this extension range may be taken as the extension range of comprehension of features of 'one'.
11. The other way to initiate oneself for enrichment of comprehension range of features of 'one' would be to reach at the different applied values of 'ONE' as a number. Illustratively, the initial range in the direction may be '1, 1^2, 1^3, and so on.
12. From here, one may jump for one unit of length, one unit of area, one unit of volume; one variable, one pair of variable, one set of triple variables; first degree of a variable, second degree of a variable, third degree of a variable, and so on.
13. Further, one may extend ones comprehension domain of one's features of 'ONE', by proceeding with '1 as 1', '2 as 1', '3 as 1', and so on.
14. This way, this skill of learning of features of 'one', may in terms of Ganita Sutras system, is to take to the next sequential step of learning skills, being 'one more' is to be learnt, and with its learning, is to be taken having attained an another progress step. Like that, one may have ones own chase of the progress of ones learning, and the skill to have

sequential self evaluation thereof would certainly being of a great help for attaining the aim of perfection of one's intelligence.

■

3

SYMMETRY INSIGHT

1. Second skill, about which the young minds are to be initiated is of 'SYMMETRY INSIGHT', as is at the base of the functional rule of Ganita Upsutra-1 'Anurupyena / to follow the proportionate symmetry of form within frames'.
2. For it the best help which the 'mind' may have is of 'body within which the mind itself is seated', and further as it is 'full of the feature of symmetry', intimacy with which to provide required 'insight' for this feature and its values.
3. Young minds may be naturally initiated for 'symmetry insight' by bringing it to their pointed attention as that their body, on its face front is accepting 'nose line as a symmetry line, while on its back, the backbone is sustaining the symmetry line.
4. The left hand and right hand as a reflection pair situations, and like that of a pair of eyes, and of a pair of ears, as well as for a pair of nostrils, and for a pair of legs, as such, illustrate the functional

feature of 'symmetry replicating the set ups in pairs, to be designated as 'mirror reflection pairs'.

5. The conceptual presence of mirror reflection pairs may be well illustrated in terms of the double digit numbers 01 to 99 splitting themselves as follows

Tables of ten-place value format

01	02	03	04	05	06	07	08	09
10	11	12	13	14	15	16	17	18
19	20	21	22	23	24	25	26	27
28	29	30	31	32	33	34	35	36
37	38	39	40	41	42	43	44	45
46	47	48	49	50	51	52	53	54
55	56	57	58	59	60	61	62	63
64	65	66	67	68	69	70	71	72
73	74	75	76	77	78	79	80	81
82	83	84	85	86	87	88	89	90
91	92	93	94	95	96	97	98	99

Tables of ten-place value format

6. It would be a blissful exercise to chase the above split of numbers 01 to 99 along the mirror lines as symmetry lines.

   Of these the central symmetry line run through the artifices 10, 20, 30, 40, 50, 60, 70, 80 and 90.

   One may, by definition accept (01, 10) as a reflection pair of artifice as these are accepting replacement for the places of digit of these numbers. Likewise would follow the reflection pairs (02, 20), (03, 30), (04, 40), (05, 50), (06, 60), (07, 70), (08, 80), (09, 90).

   Here, one shall have an insight as to the fact as that the single digit numerals (1, 2, 3, 4, 5, 6, 7,

8, 9) with their re-expressions as double digit numbers as 01, 02, 03, 04, 05, 06, 07, 08, 09, these have acquired a new feature of being reflection pairs numbers of (10, 20, 30, 40, 50, 60, 70, 80, 90).

Further it may be comprehended as that the above arrangement of numbers 01 to 99 as 9 columns and 11 rows is of the organization format which splits itself into upper and lower part such that the upper part coordinates the artifices of numbers around the mirror line running through (11, 22, 33, 44) as a reflection pairs. And further the artifices of numbers around the mirror lines running through (55, 66, 77, 88, 99) as a reflection pairs of numbers.

One shall enlist the reflections pairs of the upper part, as well as of the lower part.

One shall except the artifices 11, 22, 33 and 44 which together are constituted the mirror line, as self reflecting numbers as much as that here in such cases both the digits of the numbers are of equal values and as such it may be taken that these values of pairing with themselves as their own images. Likewise would be the position in respect of the artifices 55, 66, 77, 88 and 99.

The teachers shall help the young minds to have further insight about this coordination arrangement of the artifices of numbers of double digit of ten place value system.

7. Here the teachers may avail an opportunity to initiate young minds for transition from ten place value system to other lower place value system

and to enrich their comprehension domains about the symmetry insight in respect of the formats of place value system 9, 8, 7, 6, 5, 4, 3, 2, 1with the help of following coordination arrangements of double digit numbers of these place value system.

8. The coordination arrangement of artifices of double digit numbers of nine place value system, as under, may be availed for their chase of reflection pairs organization format:

Tables of nine-place value format

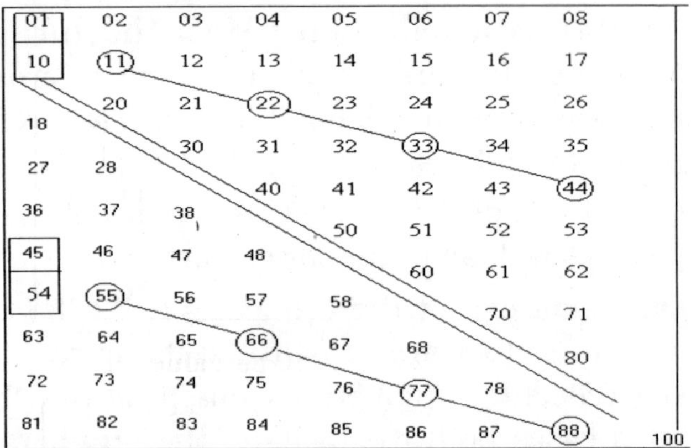

01	02	03	04	05	06	07	08	
10	11	12	13	14	15	16	17	
	20	21	22	23	24	25	26	
18		30	31	32	33	34	35	
27	28		40	41	42	43	44	
36	37	38		50	51	52	53	
45	46	47	48		60	61	62	
54	55	56	57	58		70	71	
63	64	65	66	67	68		80	
72	73	74	75	76	77	78		
81	82	83	84	85	86	87	88	100

9. Likewise, one may chase the organization formats for other place values, with their coordination arrangements of artifices for double digit being as follows:

Table-eight place value

01	02	03	04	05	06	07	
10	11	12	13	14	15	16	
	20	21	22	23	24	25	
17		30	31	32	33	34	
26	27		40	41	42	43	
35	36	37		50	51	52	
44	45	46	47		60	61	
53	54	55	56	57		70	
62	63	64	65	66	67		
71	72	73	74	75	76	77	100

Table of seven-place value

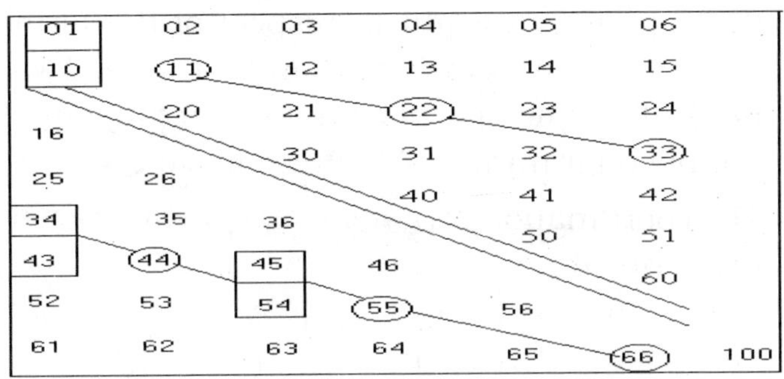

Table of six-place value

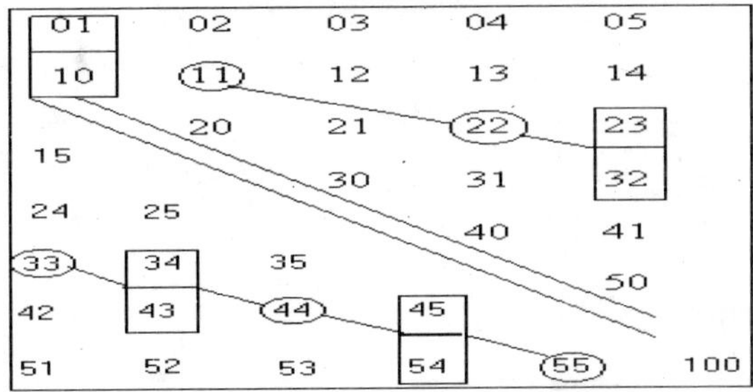

Table of five place value

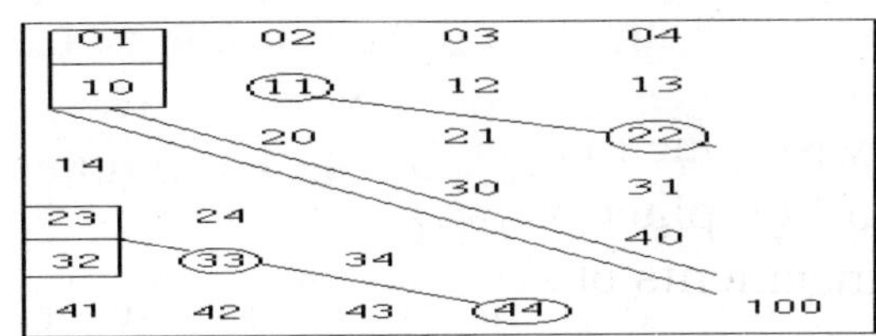

Table of four place value

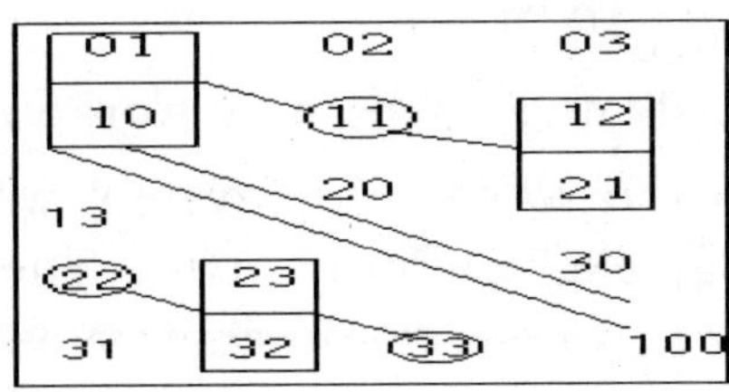

Table of three place value

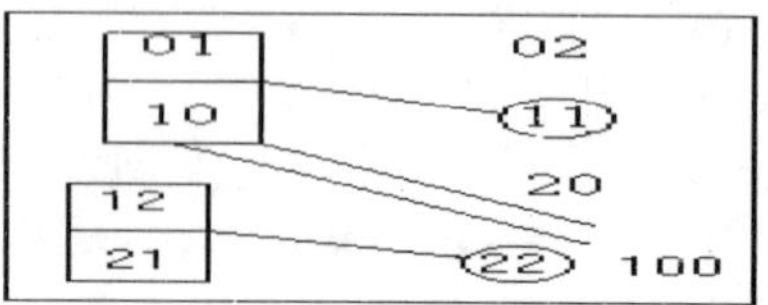

Table of two place value

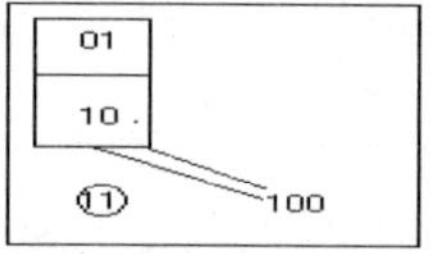

Table of one place value

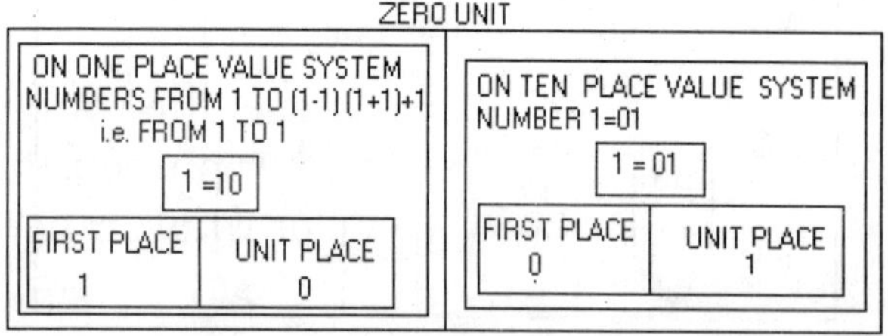

10. Teachers may aim opportunity at this phase and stage of learning to put the students to exercises of firstly asking them to ride the values of particulars numbers, say eleven in different place value systems, say 10, 9.

    As a second step the students may be asked to compute the value of double digit expression say 11 as an expression under different place value system, say 10, 9.

    These illustrative cases may workout as follows:

    Eleven in ten place value system would accept a double digit expression 11=10+1. The same in nine place value system would emerge as 12=9+2.

The second step exercise illustrations mentioned above, on its chase will take the expression 11 in ten place value system as of 'eleven' being ten plus one, and the same in nine place value system would compute to be 'ten being nine plus one'.

Like that the students may be put to large number of situations for firm comprehension of the concept and its functional rule.

11. The working rule of Ganita Sutra-1 'one more than before', works out the counts as a set of 'counting numbers / natural numbers / whole numbers / integers', while the working rule of Ganita Sutra-2 of proportionate symmetry would be taking us to the skill of 'reaching at tables'.
12. As such, even at this early stage, the young minds may be helped to have a firm grip upon the art of writing tables from 1 to 100 for whole range of 100 steps.
13. For this, the students may revisit Book-2 Section-3 TABLES 100 × 100 pages 67-72***.
14. Further for chase of tables of numerals one, two, three, four and five, six, seven, eight and nine, it may be referred to 'Book Practice Vedic mathematics (Skills for perfection of intelligence) 5.3.7 and 5.3.8 pages 207-217.***
15. This functional rule of Ganita Upsutra-1 brings to focus a 'mirror' as a mathematical tool which is to help chase symmetry and also connected range of concepts, in respect of which, more focus and attention would follow.

■

4

PLACE VALUE FEATURE

1. The transition from the organization format and functional rule of Ganita Sutra-1 (एकाधिकेन पूर्वेण *Ekadhikena Purvena* / one more than before) to the organization format and functional rule of Ganita Sutra-2 (निखिलं नवतष्चरमं दषत: *Nikhilam Navatascaramam* / All from 9 and the last from ten) takes from the 'counting based skill' to 'place value based skill with focus upon ten place value system.
2. The place value based skill, as the formats suggests, the same focuses upon 'place value' which here is 'ten place value'.
3. This focus is evidently concentrating upon the shift for a counting step ahead of 'nine', as much as that up till counts 1 to 9, these remain numerals while a step ahead there of avail a format of distinct features than that of numerals.
4. The teachers may avail an opportunity at this phase and stage of learning to focus attention of young minds as to the numerals being 'single digits expressions', while ahead of them would follow double and multiple digit numbers.

5. The numerals one to nine be taken as constituting a class in itself. The first feature of them being that these all are 'of single digit expressions'. Secondly, as that these may be taken as of 'linear format' despite these having being part of the place value system. Thirdly that, these all are to have a placement in the system at the start with place itself. Moreover, this start with place in the place value system is of affine state / zero degree power, which in fact makes all of them (One to nine), with zero power expression, is to take them to 'unit count' place.
6. Further the teacher, may also availed this phase and stage of learning of this skill of place value features, to impress upon the ordering of the place values as increasing in the sequential order of degrees of power of place value itself as zeroth degree power, first degree power, second degree power and so on, viz, $(10)^0$, $(10)^1$, $(10)^2$, and so on.
7. Here itself the young minds may be exposed to different place values systems by firstly having an expression for the above sequential order $(10)^0$, $(10)^1$, $(10)^2$, and so on as X^0, X^1, X^2, and so on. Then this general variable (X) may be associated the values of the concerned place value system. Illustratively if the working is in terms of nine place value system then the sequence X^0, X^1, X^2, and so on would emerge to be as 9^0, 9^1, 9^2, and so on. Likewise to follow the systems for other place values orders.
8. Further here, the opportunity may be availed to expose the young minds to the property of domain-boundaries ratios of interval, square, cube and

hyper cubes as providing formats for the sequential orders of place values systems. In particular, the exposure to ten place value system may be impressed upon along the format of hyper cube 5 in terms of its domain value ratio being $A^5:10\ B^4$.

9. The features, firstly as that all powers of one are of value one itself and secondly as that the origin of hyper cube 4 is of a transcendental core as the same being the seat of transcendental worlds (5-space / hyper cube 5) and thereby from within the origin of each of the ten hyper cubes 4 would ascend 10 hyper cubes 4 enveloping hyper cube 5, and thereby there would emerge as many as 10 x 10 = 100 = 10^2 hyper cube 4 and enveloping boundary of boundary of transcendental worlds (5-space). And this process of ascendance, at each sequential step, would go on multiply by 10 and thereby, there being a sequential expansion of the order of degrees of powers of ten place value system, namely, 10^0, at the initial affine state of transcendental worlds, followed by the sequential steps of degrees of powers of order 10^0, 10^1, 10^2, 10^3, and so on.

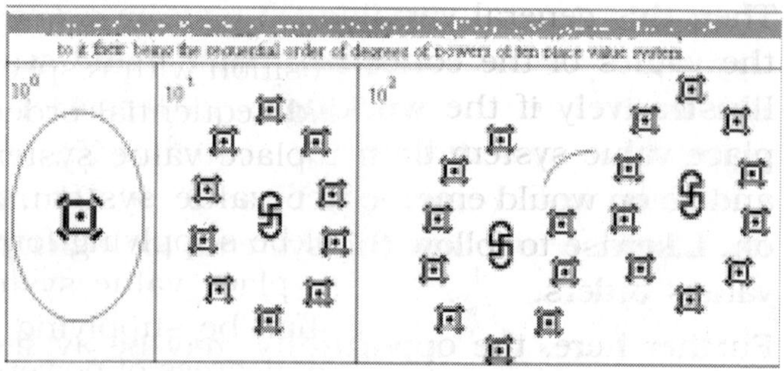

10. Further the opportunity may be availed by the

teachers here at this phase and stage of learning of other even place values, say, 'eight' availing format of hyper cube 4, and likewise eight place value system is to avail the format of hyper cube 4.

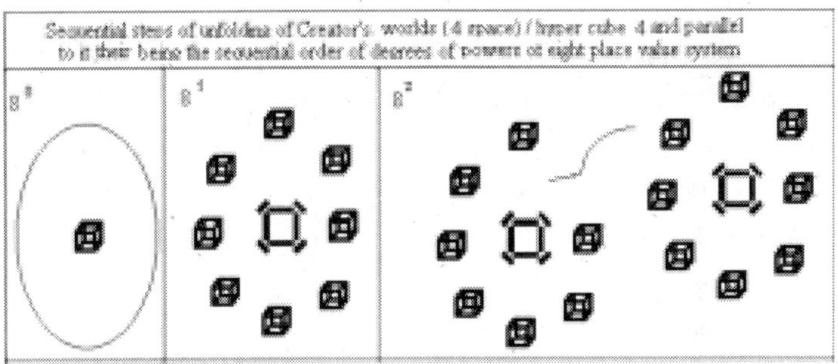

11. Here, it may be impressed upon the young minds as that for odd place value system, the formats of circle, sphere and hyper spheres are to be availed, and the same are to be chased at next phase and stage of learning at high and secondary schooling.

    Still further, here itself an opportunity may be availed to expose the young minds as that the nine versions of hyper cube 4 shall be supplying the formats for nine numerals of ten place value system while format of hyper cube 5 itself shall be supplying the format for the sequential order of degrees of powers of ten place value system. And, parallel to it would be the position with respect to the formats of numerals and sequential order of degrees of powers of other place value systems. Illustratively for eight-place value system, the seven versions of cube shall be supplying format for seven numerals of eight-place value system, while hyper cube 4 itself shall be supplying the format for sequential order of degrees of powers of eight place value systems. ■

SECTION - 3
CHASE OF SET UP OF A CUBE

1
INTRODUCTORY

For proper foundation of chase of ancient wisdom of range 'arithmetic to astronomy', appropriate conceptual foundation is to be laid for it beginning with the comprehension of the set up of the cube in a static as well as in a dynamic position.

For it, at initial stage, at which at present the young minds are to be exposed, this may be approached in two ways, firstly while sitting within a room (cube), and secondly while viewing the cube from outside.

The comprehension of the set up of a cube from its within, be focused taking as that the cube is in a static position.. the comprehension of the set up of the cube from outside of it may be focused taking as that the cube is, 'as it is', as a set up enveloped within a synthetic boundary as a package, and also as it is to be while it is in dynamic state under different situations.

There may be a different set of situations, prominently amongst them to be as that cube is in its biggest form in a sphere. Parallel to it may be a situation as it containing a biggest sphere within it.

Then the connected pair of cubes in such situations may pose a chase situation which deserves to be learnt to be chased for perfection of intelligence as to the Reality of the Creator's space (4-space) as well as of transcendental worlds (5-space), and self referral domains (6-space) and the sky in between the range of 'Earth to Sun'.

■

2

SETUP OF A CUBE

One may evaluate one's learning in terms of one's knowledge of the set up of a cube.

For acquiring knowledge about the set up of the cube, to start with, there may be two distinct approaches, as to be of viewing from outside and as to be knowing from within the cube.

Viewing from outside, would result into the knowledge of the geometric envelope as synthetic boundary for the cube as a domain.

The knowledge from within, is to be of the view as to how the volume as an expression of the domain content is sustaining its center as a seat of origin of 3-space with 4-space and whole range of origins of spaces standing compactified here.

The comprehension of the set up of the cube shall be focusing about its dimensional order, boundary, volume and the center, as folds of distinct roles of distinct spaces, namely line / 1-space in the role of dimension, plane / 2-space in the role of boundary, the solid domain because of 3-space itself and the 4-space in the role of the origin fold of this set up.

The synthetic set up of the boundary as geometric envelope stitched availing geometric entities namely corner points, edges and surface plates.

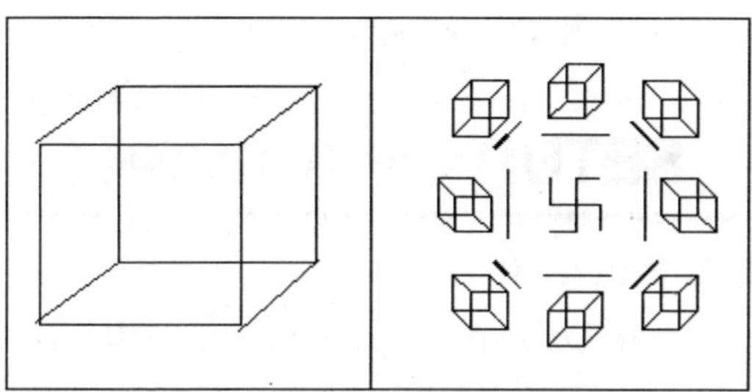

TRANSITION FROM CUBE TO HYPER CUBE-4

1. The set up of cube as representative regular body of 3-space be taken as the set up of a linear order mental format of waking state of consciousness state as old mental format.
2. The set up of hyper cube 4 as representative regular body of 4-space be taken as the set up of a spatial order mental format of dream state of consciousness state as new mental format.
3. The sadkhas fulfilled with an intensity of urge to fully know and to chase the transition process for its old mental format to that of new mental format, shall first start learning the features of the set up of the cube.
4. One shall chase these features of the set up of the cube at intellectual format as well as at experimental glimpsing state of the transcending mind.
5. One shall transcend through the synthetic boundary of the cube and reach at its domain.

6. One shall further transcend through the domain of the cube and reach at its center.
7. Still further one shall permit the mind to transcend and to un lock the seal at the center of the cube.
8. One shall continue in prolonged sittings of trans till the seal at the center of the cube / origin of 3-space gives way to the transcending mind into the Creator's space / 4-space.
9. One shall continue transcending till one is of the privilege of glimpsing the wide open domain of Creator's space.
10. With center of cube / origin of 3-space giving way to the transcending mind and the wide open domain of Creator's space well coming the glimpsing mind, a process of transition for the old linear order format of the transcending mind shall take place and in the process the degree of freedom of motion of the cube within Creator's space because of the additional dimension being available, the cube shall be manifesting and additional edge, to be designated as the 13th edge of the cube.
11. The additional edge of the cube shall be availing the additional dimension which would be of a spatial order, and with it, in the process of transcendence and glimpsing of the Creator's space, the transcending mind shall be acquiring transition for its old linear order mental format into a new spatial order mental format parallel to the emergence of spatial format for the additional edge of the cube within hyper cube 4 / 4-space domain / Creator's space.

12. With attainment of a spatial expression format for an additional edge of the cube within hyper cube 4 along its fourth dimension, a process for transition and transformation for the three linear dimensions of a cube / 3-space into three spatial dimensions of hyper cube 4 / 4-space shall be in the process of attainment.
13. It is this attainment of spatial order and also the attainment of an additional dimension which results into the transformation for a 3-space setup into a 4-space set up of four spatial dimensions.
14. It is this set up of four spatial dimension which require a set of eight points for fixation of a point within the hyper cube 4.
15. The center of the cube as a point expression of hyper cube 4, with opening of its seal, shall be splitting the cube into 8 sub cubes and a 3-space format into 8 octants, and in the process the eight corner points of eight sub cubes of the cube shall be unfolding the seats of eight solid boundary components of hyper cube 4.
16. With this transition for the linear order into a spatial order shall be setting into motion a process of transformation during which the point value hyper cube 4 at center of the cube shall be sequentially evolving into hyper cube 4 of higher values, and ultimately as a full 4-space domain.
17. One shall chase this evolutionary process of the range beginning with zero value hyper cube at center of the cube and at the other end the wide open Creator's space / 4-space domain.
18. One may also intellectually chase the set up of

hyper cube 4 as of domain boundary ratio A^4: 8 B^3 as a fourth member of the sequence of hyper cube 4, that is A^N: 2NB^{N-1}, N=1, 2, 3, 4 —.

19. Teachers shall also help the students to chase hyper cube 4 as a manifestation of four folds with 2-space in the role of dimension, 3-space in the role of boundary, 4-space as domain and 5-space as origin.

■

3

FLOW ALONG EDGES

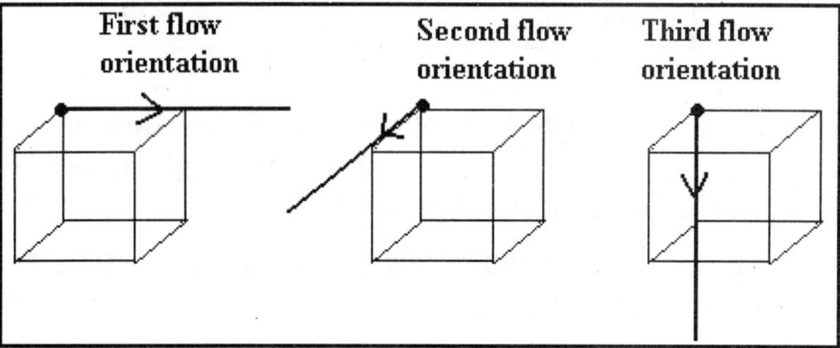

1. One shall see that three edges meet at each corner point of the cube.
2. One may view it as a feature of this set up as that from each corner of a cube their may be three distinct flow orientations along three distinct edges.

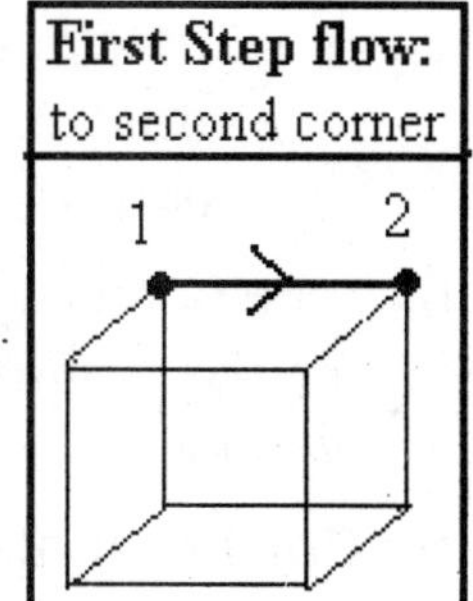

3. It would be taken as an exercise to carry forward along the orientation of a given edge

to connect all the eight corner points.

4. Here, one may be confronted with a pair of options at the second step, as is being illustrated here under.

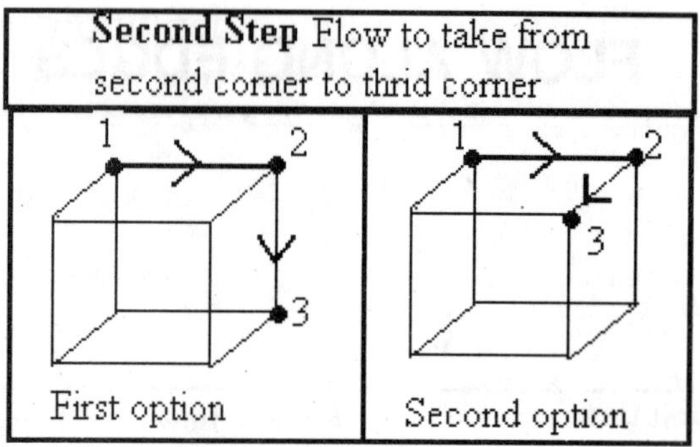

Step 1. To proceed along the first orientation and reach the other end of the edge at second corner of the Cube.

Step 2. At the second corner, the flow to have two options as under:

Step3. Either of the options, while reaching third corner, shall be again confronted with a pair of options. The same be followed individually. And, at next step as well there to follow a pair of options, but with that, the rest of the flow is to have only single flow option to connect remaining all the corner points.

5. It be taken, as an exercise to workout all the options for the original first orientation flow from the first corner.

6. Further it also be taken as an exercise to workout

the flow options for the second and third flow orientations from the first corner.

7. Still further, it also be taken as an exercise to workout the similar flow options for all the eight corners of the cube.
8. Still further, as the flow orientation takes from first corner to the 8th corner in seven steps availing coordination arrangement of seven edges in a sequence, and as each such flow may be taken in reverse orientation, as starting with the 8th corner and reaching up till the first corner, as such the question would stand posed for an answer as to how many pairs of such orientations are accepted by the set up of the cube. Before actual chase and working out of all the orientation flow coordination of all the 8th corner at a time, one may attempt the poser as to, if the total pairs of orientations taking from first corner to eighth corner, and in reverse orientation from eighth corner to first corner, to be precisely 8 x 3 =24.
9. One may pause and have a fresh look at the set up of the cube and to chase and comprehend on the format of the flow orientation connecting a pair of corner points in terms of a connecting edge, as that the edge, as a line is a flow of a moving corner / point.
10. One may attempt definition of a line as a flow track of a moving point.
11. One may attempt chase path for a flow track along the format of a line in a pair of ways parallel to the pair of the orientations of a line.

12. One may attempt to define a direction for the other end corner point in terms of the orientation of the edge leading from first corner point to the second corner point.

■

4

SPLIT STREAMS

1. The students shall chase the possible number of ways all the eight corner points may be coordinated through edges of the cube by maintaining the orientation and without visiting any of a corner points more than once.

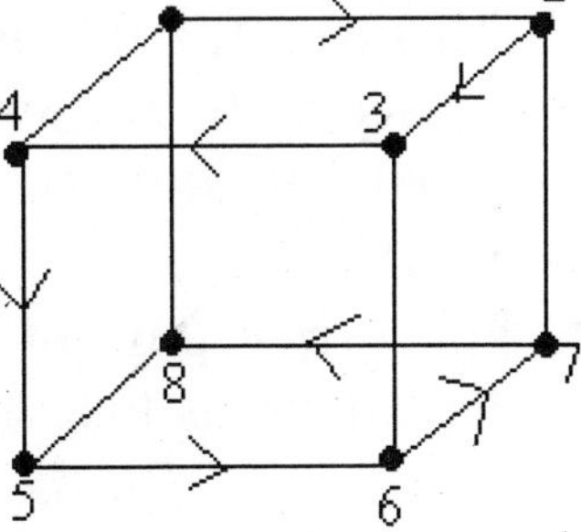

2. Let us fix our starting point to be corner '1'.
3. Let us fix our orientation for flow at the initial stage of corner '1' to be from corner '1' to corner '2'.
4. This orientation flow shall be taking us through seven steps connecting

all the eight corner in a sequence (1, 2, 3, 4, 5, 6, 7, 8, 9).

5. We can see that at corner '2', the split for the flow could take along corner '3' or towards corner '7'. The split flow which take from corner '2' towards corner '3' is a flow as has been chased above as (1, 2, 3, 4, 5, 6, 7, 8, 9).

(1, 2, 7, 8, 5, 4, 3, 6)

The split flow which is to take from corner '2' to corner '7' is of chase steps. (1, 2, 7, 8, 5, 4, 3, 6).

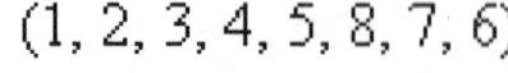

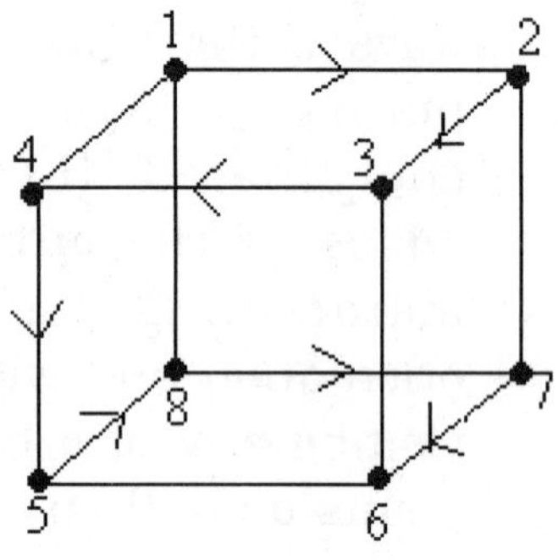

6. The flow which has already taken us up till corners 1, 2, 3, 4 and 5, has a split for corner 'eight', as well as for corner '6'. The first flow stream of this split is (1, 2, 3, 4, 5, 6, 7, 8) which has been depicted above in para 4. The second split stream is (1, 2, 3, 4, 5, 8, 7, 6).

7. The other pair of flow streams would permit depictions as follows

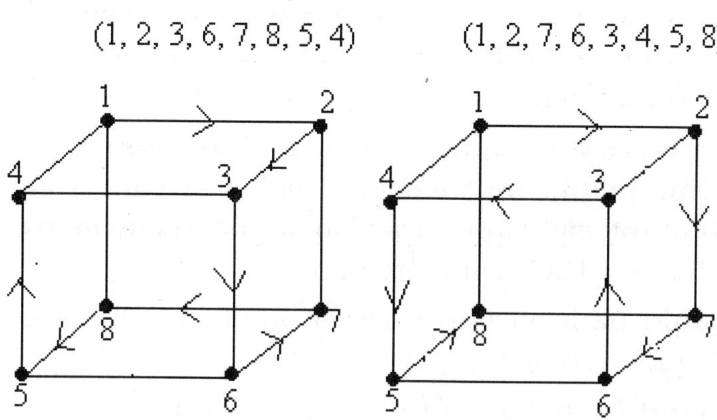

8. This way, it may be comprehended and glimpsed as that the flow starting from corner '1' with orientation along edge leading to corner '2' results into 'five distinct flow streams of seven steps each coordinating eight corner in the following sequences.

Stream	Flow sequence	Flow depiction
1	Corners 1, 2, 3, 4, 5, 6, 7, 8	(1, 2, 3, 4, 5, 6, 7, 8)
2	Corners 1, 2, 7, 8, 5, 4, 3, 6	(1, 2, 7, 8, 5, 4, 3, 6)
3	Corners 1, 2, 3, 4, 5, 8, 7, 6	(1, 2, 3, 4, 5, 8, 7, 6)
4	Corners 1, 2, 3, 6, 7, 8, 5, 4	(1, 2, 3, 6, 7, 8, 5, 4)
5	Corners 1, 2, 7, 6, 3, 4, 5, 8	(1, 2, 7, 6, 3, 4, 5, 8)

9. Now from corner '1' there can be a flow orientation towards '4' or even towards corner '8'.

10. Accordingly, it may be accepted that five flow streams would be available for each of these two orientation as well, and as such, in all there are to follow as many as 5 x 3 = 15 flow streams.
11. Then there are eight corners, and as such eight starting point, and hence 15 splits stream from each of the eight corner are to supply us as many as 15 x 8 = 120 splits streams.
12. It would be a blissful exercise to chase all these 120 splits streams.
13. It would be helpful to have a sequence of eight points along the circumference of a circle (1, 2, 3, 4, 5, 6, 7, 8). Being a sequential placement along the circumference it shall be providing a privilege to start from any of these eight placements and to cover the whole range of sequential flow.
14. This shall be yielding '8' settings as follows.

Setting	Sequential order
1	(1, 2, 3, 4, 5, 6, 7, 8)
2	(2, 3, 4, 5, 6, 7, 8, 1)
3	(3, 4, 5, 6, 7, 8, 1, 2)
4	(4, 5, 6, 7, 8, 1, 2, 3)
5	(5, 6, 7, 8, 1, 2, 3, 4)
6	(6, 7, 8, 1, 2, 3, 4, 5)
7	(7, 8, 1, 2, 3, 4, 5, 6)
8	(8, 1, 2, 3, 4, 5, 6, 7)

15. It would be interesting exercise to shift from five split flow streams of orientation (corner 1 to corner 2) of setting sequence (1, 2, 3, 4, 5, 6, 7, 8) to parallel five fold split flow streams for other seven

settings by simply following the circular sequence order of the settings as follows:

(i) A shift from setting 1 to setting 2.

Stream	Setting 1	Setting 2
1	(1, 2, 3, 4, 5, 6, 7, 8)	(2, 3, 4, 5, 6, 7, 8, 1)
2	(1, 2, 7, 8, 5, 4, 3, 6)	(2, 3, 8, 1, 6, 5, 4, 7)
3	(1, 2, 3, 4, 5, 8, 7, 6)	(2, 3, 4, 5, 6, 1, 8, 7)
4	(1, 2, 3, 6, 7, 8, 5, 4)	(2, 3, 4, 7, 8, 1, 6, 5)
5	(1, 2, 7, 6, 3, 4, 5, 8)	(2, 3, 8, 7, 4, 5, 6, 1)

Note, All what has been done is that 'one more count' is added at each of the steps. The corners 1, 2, 3, 4, 5, 6, 7, 8 have been changed into 2, 3, 4, 5, 6, 7, 8, 1.

(ii) Likewise there can be a shift from five fold streams of corner 2 to five fold streams of corner 3. And likewise the process to continue up till a shift from five fold streams of corner seven to five folds streams of corner 8.

(iii) Taking this orientation of corner '1' to corner '2' as orientation of x axis (first axis), there may be a parallel depiction for orientations along y axis and z axis of a three dimensional frame embedded into corner '1'. The students may take the help of the teachers in chase of all these 120 splits stream flow along the edges of the cube. ■

5

BACK TO THE STARTING POINT

1. The seven steps long flow streams running through all the eight corners of the cube, can take back to the starting position but it would amount to re visiting the start with corner.
2. This way the spatial boundary of the cube shall be having as many as 120 circular flow paths and these together shall be making the spatial boundary of the cube as to be of very rich potentialities which deserve to be chased for harnessing its values.

■

6

BACK TO THE SOURCE RESERVOIR

1. The seven steps long flow streams, instead of re-visiting the start with corner, may get connected with the center / origin as the source reservoir.
2. The flow from the source reservoir may reach any of the corners, and its seven steps long flow stream running through all the eight corners shall be taking back to the source reservoir.
3. This way, as many as 120 flow streams originating from the source reservoir / center / origin and after flowing through all the corner reaching back the source reservoir, is a phenomenon which deserves to be chased.
4. This phenomenon of 120 artifice is structurally very rich set up and is of the order of 120 hyper cubes 4 coverage of boundary of boundary of hyper cube 6.

■

7
TRANSCENDENCE FROM WITHIN

1. This phenomenon of 120 flow streams originating from the source reservoir / center / origin and after running through all the corners and reaching back the source reservoir is a transcendental phenomenon of transcendence from within the center of the cube / origin of 3-space.
2. This transcendental phenomenon of transcendence from within the center of cube / origin of 3-space as seat of hyper cube 4 / 4-space / Creator's space deserves to be comprehended intellectually as well as the same deserves to be experientially glimpsed while permitting the mind to transcend through the spatial boundary of the cube and to reach the center of the cube / origin of 3-space / seat of 4-space and to continue transcending till the new wonderful transcendental worlds are glimpsed.

Split up of a cube as one thousand sub cubes

The blissful exercise to appreciate and to have an insight into the set up of a cube and to have perfection of intelligence in terms of understanding of the setup, one may chase an organization of a cube as synthetic set up of one thousand cubes.

This chase may help appreciate the Vedic wisdom of chase the boundary of hyper cube 5 in terms of ten place value system for transition and transformation for the transcendental carriers carrying solid order to hyper solid self referral domains (6-space).

This further would provide insight as to the transcendental carriers within rays of the sun sustaining and carrying ***Triloki*** /Vishwa / Jagat / 3-space / cube.

This set up may be chased as under

Step 1 take a cube of edges of ten units.

Step 2 this cube would be equal to 1000 cubes.

Step3 these 1000 cubes can be taken being arranged as 10 slabs of 100 cubes each.

Step 4 We may first take up the upper slab of 100 cubes.

Step5 This slab may be taken as arranged as 10 rows of 10 cubes each.

Step 6 Of these, we may begin with its first row of 10 cubes.

Step 7 of these 10 cubes may start with the first cube of the row.

Step 8 this first cube of first row of first slab shall be supplying 8 corner points ,12 edges, and 6 surfaces.

Step 9 the other nine cubes of this row shall each being supplying 4 corners, 8 edges and 5 surfaces.

Step 10 This way the contribution of these 9 cubes

of first row would be 4 × 9 = 36 corners, 8 × 9 =72 edges and 9 × 5 = 45 surfaces.

Step 11 As such, the contribution of all the ten cubes of first row of first slab would be 36+8=44 corners, 72 +12 = 84 edges and 45 +6= 51 surfaces.

Step 12 Now, the second row of the first slab may be taken up for its contribution.

Step 13. The first cube of the second row shall be contributing 4 corners, 8 edges and 5 surfaces.

Step 14 Each of the other 9 cubes of the second row shall be contributing 2 corners, 5 edges and 4 surfaces each. This way the total contribution of all the nine cubes would be 18 corners, 45 edges and 36 surfaces.

Step 15 this way the total contribution of the second row of first slab would be as of 22 corners, 53 edges and 41 surfaces.

Step 16 As the contribution for second to tenth rows of the first slab is to be identical, therefore the total contribution of these 9 rows of the first slab would be 22 × 9 = 198 corners, 53 × 9 = 477 edges and 41 x 9 =369 surfaces.

Step 17 This take us to the total contribution of the first slab as to be of its all the ten rows, comes to be as of 44+198=242 corners, 84 +477=561 edges and 51 +369=420 surfaces.

Step 18 Now, the second slab may be taken up for its contribution.

Step 19 First of all the its first row is taken up.

Step 20 Of this first row, as well the first cube is taken up.

Step 21 The contribution of first cube would be of 4 corners, 8 edges and 5 surfaces.

Step 22 Then the contribution each of the remaining 9 cube of this row would be as of 2 corners, 5 edges and 4 surfaces. AS such the total contribution of all 9 cubes as of 18 corners, 45 edges and 36 surfaces.

Step 24 With this the total contribution of all the ten cubes of first row of second slab would be as of 22 corners, 53 edges and 41 surfaces.

Step 25 The contribution of second to 9th rows of the second slab would be equal

Step 26 The contribution of the first cube of second row of second slab would be as of 2 corners, 5 edges, and 4 surfaces.

Step 27 The contribution of second cube of second row would be as of 1 corners, 3 edges and 3 surfaces.

Step 28 Therefore the total contribution of second to 10th cube of the second row would be as of 9 corners, 27 edges and 27 surfaces.

Step 29 As such the total contribution of the second row of the second slab 2+9=11 corners, 5 +27=32 edges and 4+27=31 surfaces.

Step 30 As the contribution for second to 10th row of the second slab is to be equal, as such the total contribution of these 9 rows, namely second to 10th row of second slab would be as of 11 x 9 = 99 corners,32 × 9=288 edges and 31 x 9 =279 surfaces.

Step 31 Therefore the total contribution of the second slab of the second row would be of 22+99 =121 corners, 53+288 =341 edges and 41+279 = 320 surfaces.

Step 32 Now as the contribution for second to tenth

slabs is to be identical, as such the total contribution for these 9 slabs namely of second to 10th slabs would be as of 121 x 9=1089 corners, 341 x 9 =3069 edges and 320 x 9=2880 surfaces.

Step 33 This would lead us to total contribution of all the ten slabs as to be of 242+1089 =1331 corners, 561+3069=3630 edges and 420+2880 =3300 surfaces

Step 34 The total entities contribution of the entire set up comes to be 1331+3630+3300= 8261

Step 35 The chase on the format of corner points / points / 0-space set up would be as of artifice of 1331. one may see that it amounts to approaching from either side of the interval in the sequential order of artifice 1 leading to artifice 3. parallel to it would be 1-space leading to 3-space. This would, as such give us an insight as when we proceed the set up of a cube in terms of half dimensions, it gives only a chase up till the middle but for the middle point. This way even the chase from either end of the edge, is to leave the middle / center / origin as to be an isolated point and it is to remain envelope within the soild boundary.

Step 36 Now when the set up of cube is chased on the format of a line / edges, it leads to the set up of the cube as of the order of the artifice '3630'. The edges here being of ten units each, so for the single unit, the contribution would emerge as to be of the order of artifice 363. This chase when is to be had in terms of the Devnagri numerals, one can see that these numerals three and six being a reflection pair and the arrangement of the artifice 363 is that the middle is reached by the light in terms of its reflection operation and makes the middle joint as of double count. It is in this context that it would be relevant to have a focus attention upon the number value format of joint, which comes to be NVF (JOINT) = 68 = 34 +34 = NVF (ONE) +NVF (ONE). With it it may be a phase and stage to impress upon the young minds for their focus attention to the fact as that the center of the cube / origin of 3-space is of a spatial order while the domain of the cube remains to be of a linear order.

Step 37 Further, when the set up of the cube is chased on the format of a surface, it leads the order of the set up of the cube as to be of the order of artifice 3300. Here it may be relevant to note that the spatial order, in terms of the split for the axis as a reflection pair (01,10) shall be acquiring coordination arrangements for the organization format as (01 x 01, 10 x 10). With it the artifice 3300, shall be providing '33' dimensional units for ascendance. This artifice (33) is of the generic order of the number value format of 'seed' , as that nvf (seed) = 33. Here is an opportunity at this phase and stage of learning as that the split of a three dimensional frame into a pair of half dimensions frames, is the phenomenon of the set up of cube / 3-space which deserves to be chased with focus, as that it is the reversal and translation thereof from the center of the cube to the corner points, which is attained and it gives inherent feature to the setup of the format of the cube as to be of the privileged status of a representative regular body of 3-space as manifestation along the Creator's format.

Step 38 The sadkhas, for perfection of their intelligence may glimpse the set up of a cube within a Creator's space and have the bliss of the spatial order at work for making 'cube' as well as the sphere as to be of same domain boundary ratios.

Step 39 once one has comprehended these features of manifestation of the set up of a cube, one may proceed ahead for the applied values of this set up, as it being of the within a sphere or that it sustaining a sphere within itself.

Step 40 Ones, one is comfortable with the set up of a cube, (and of even place value systems), one may proceed ahead for initiations of learning steps for the set up of a pair of connected spheres (and of odd place value systems).

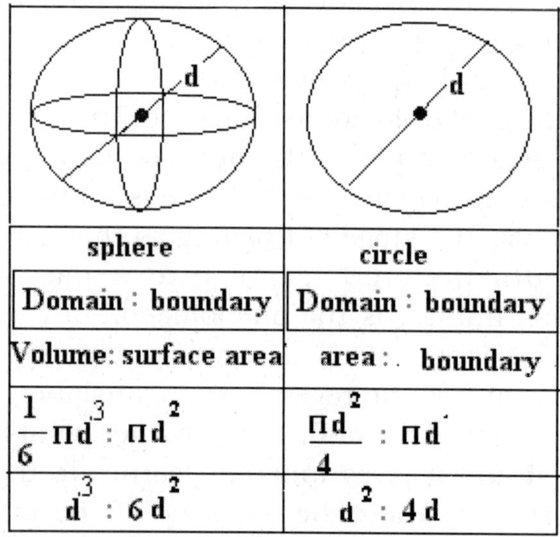

sphere	circle
Domain : boundary	Domain : boundary
Volume: surface area	area : boundary
$\frac{1}{6}\pi d^3 : \pi d^2$	$\frac{\pi d^2}{4} : \pi d$
$d^3 : 6 d^2$	$d^2 : 4 d$

Step 41 For a smooth transition from the set up of the cube to the set up of the sphere, it be phased as is to be a transition from a square to a circle, and a step ahead from cube to a sphere. The sequential order of square, cube and hyper cubes on the one hand and

the sequential order of circle, sphere and hyper spheres on the other hand, may be chased as parallel orders, and in the process, the smooth transition from the set up of the cube to the set up of the sphere may be attained. Once it is attained, for which the following steps may be helpful.

■

SECTION - 4

CHASE ON NUMBER VALUE FORMATS OF WORDS FORMULATIONS

1

INTRODUCTORY

1. Orthodox English vocabulary is one such source of words formulations which avail number value formats.
2. Chase of number value formats of these formulations, is bound to be of rich dividends. Firstly it shall give us insight about the number value formats as a technique of organization of knowledge. Secondly, it shall be providing us very rich source material for chase of partitions and coordination arrangements of artifices of numbers. Illustratively NVF (TRUTH) = 87, and corresponding to partition to 87, (2+85), ——(43, 44), and further the such like partitions for whole range of numbers 1 to 86, shall be yielding a very rich spectrum of inner partitions and coordination of artifice 87, and the same when would be chase in terms of the specific words formulations of such number value formats corresponding the partition values, these shall be enriching with the whole range of features of the organization formats. Illustratively, the partition (43, 44) shall be of the order of NVF (SPACE FRAME). Likewise the partition (50, 37), shall be of the order of NVF (VOID CONE). ■

2
DEFINITIONS: NVFS

NVFs (number value formats) of alphabet letters It be accepted by definition as that:

NVF (A)=1	NVF (B)=2	NVF (C)=3	NVF (D)=4
NVF (E)=5	NVF (F)=6	NVF (G)=7	NVF (H)=8
NVF (I)=9	NVF (J)=10	NVF (K)=11	NVF (L)=12
NVF (M)=13	NVF (N)=14	NVF (O)=15	NVF (P)=16
NVF (Q)=17	NVF (R)=18	NVF (S)=19	NVF (T)=20
NVF (U)=21	NVF (V)=22	NVF (W)=23	NVF (X)=24
NVF (Y)=25	NVF (Z)= 26		

NVFS OF WORDS

It be accepted as definition as that NVF of given word is sum of the NVFs of the alphabet letters availed for the composition of the given words.

ILLUSTRATIONS

1. NVF (GOD)

   =NVF(G)+ NVF (O)+ NVF (D)

   =7+15+4

=26

=NVF(Z)

2. NVF (WORD)

=NVF(W)+NVF(O)+NVF(R)+NVF(D)

=23+15+18+4

=60

=NVF(FOUR)

3. NVF (ENGLISH)

=NVF(E)+ NVF (N)+ NVF (G)+

NVF (L)+NVF (I)+NVF (S)+NVF (H)

=5+14+7+12+9+19+8

=74

=NVF(PAIRING)

=NVF(BIBLE SPACE)

=NVF(POINT)

4. NVF(TABLES)= NVF(SOLID)=NVF(CIPHER)

NVF OF GROUP OF WORDS

It be accepted as definition that NVF of group of words is the sum of NVFs of words of the group.

Illustrations

NVF (BIBLE SPACE)

=NVF(BIBLE)+NVF(SPACE)

=30+44=74

=NVF(PARING)

=NVF(ENGLISH)

NVF (PAIRING SPACE)

=NVF(PAIRING)+NVF(SPACE)

=74+44

=118

=NVF(DICTIONARY)

NVF(NVFs)

=NVF(N)+ NVF (V)+ NVF (F)+(NVFS)

=14+22+6+19

=61

=NVF(CHURCH)

=NVF(GENERIC)

NVF(BIBLE GENERIC)

=NVF(BIBLE)+ NVF (GENERIC)

=30+61

=91

=NVF(MIRROR)

NVF(CHART)

=50

=NVF(VOID)

NVF(CHARTS)

=69

=NVF(VOIDS)

=NVF(ZOOM)

NVF(NVFs DICTIONARY CHARTS)

=NVF(CHURCH PAIRING SPACE ZOOM)

■

3

TABLE OF NVFS 1 TO 100

Here is arrived at one Table of NVFs 1 to 100 and Sadhkas may further tabulate another set of formulations of this range of NVFs 1 to 100 as well as for extended ranges and sub ranges and also to attempt chart for voids zooms in terms of settled NVFs formats permissible in terms of tabulated formulations reached at by the Sadhkas.

NVF	WORD	NVF	WORD
100	DISCIPLINE	50	VOID
99	THOUGHT	49	AXES
98	TRANSCEND	48	POLE
97	PILGRIMAGE	47	MONAD
96	KNOWLEDGE	46	LOGIC
95	RENEWING	45	RANGE
94	TRI-MONAD	44	SPACE
93	ENTITY	43	FRAME
92	REVERSE	42	NEW
91	MIRROR	41	AFFINE
90	ARTIFICES	40	LINE
89	UNITY	39	ANGLE
88	VOLUME	38	ION
87	TRUTH	37	SEAL
86	PARAGRAPH	36	REAL
85	CREATION	35	EYE
84	COLOUR	34	ONE

83	FOLLOW	33	SEED
82	FACTORS	32	LIFE
81	SQUARE	31	CUBE
80	CREATOR	30	BIBLE
79	NATURE	29	BLACK
78	AMBROSIA	28	AIR
77	MATTER	27	HALF
76	MINUS	26	Z
75	SERIES	25	Y
74	PAIRING	24	X
73	FORMAT	23	W
72	ORIGIN	22	V
71	SPHERE	21	U
70	STOP	20	T
69	ZOOM	19	S
68	JOINT	18	R
67	WATER	17	Q
66	SINGLE	16	P
65	CENTER	15	O
64	ZERO	14	N
63	LIMIT	13	M
62	COSMIC	12	L
61	CHURCH	11	K
60	FOUR	10	J
59	SOLID	9	I
58	TWO	8	H
57	HEIGHT	7	G
56	LIGHT	6	F
55	SKY	5	E
54	SUN	4	D
53	AXIS	3	C
52	EARTH	2	B
51	FULL	1	A

■

4

TABLE OF PARTITIONS OF NVF (TRUTH)= 87

TABLE OF PARTITIONS OF NVF (TRUTH)= 87

To have an idea of the spectrum of features of the formats and artifices pairing, here below is being attempted one pairing table of partitions of artifice 87 as NVF(TRUTH)

Sr.no	Partition	First Formulation	Second Formulation
1	86,01	PARA-GRAPH	A
2	85,02	CREATION	B
3	84,03	COLOUR	C
4	83,04	FOLLOW	D
5	82,05	FACTORS	E
6	81,06	SQUARE	F
7	80,07	CREATOR	G
8	79,08	NATURE	H
9	78,09	AMBROSIA	I
10	77,10	MATTER	J
11	76,11	MINUS	K
12	75,12	SERIES	L
13	74,13	PAIRING	M
14	73,14	FORMAT	N
15	72,15	ORIGIN	O
16	71,16	SPHERE	P
17	70,17	STOP	Q
18	69,18	ZOOM	R

19	68,19	JOINT	S
20	67,20	WATER	T
21	66,21	SINGLE	U
22	65,22	CENTER	V
23	64,23	ZERO	W
24	63,24	LIMIT	X
25	62,25	COSMIC	Y
26	61,26	CHURCH	Z
27	60,27	FOUR	HALF
28	59,28	SOLID	AIR
29	58,29	TWO	BLACK
30	57,30	HEIGHT	BIBLE
31	56,31	LIGHT	CUBE
32	55,32	SKY	LIFE
33	54,33	SUN	SEED
34	53,34	AXIS	ONE
35	52,35	EARTH	EYE
36	51,36	FULL	REAL
37	50,37	VOID	SEAL
38	49,38	AXES	ION
39	48,39	POLE	ANGLE
40	47,40	MONAD	LINE
41	46,41	LOGIC	AFFINE
42	45,42	RANGE	NEW
43	44,43	SPACE	FRAME

Sadkhas may attempt tabulation for above partition 1 to 43 of artifices 87 with the help of another set of formulations of NVFs 1 to 86

Sadkhas may also tabulate different formulations accepting common NVFs as 87.

■

5

TABLE OF PARTITIONS WITH 'TRUTH' AS ONE PART

Still further sadkhas may tabulate partition of different artifices with 87 as one of the artifices. One such tabulation for artifices 88 to 174 is being drawn here under.

Sr.no	Partition	First Formulation	Second Formulation
1	86,01	PARA-GRAPH	A
2	85,02	CREATION	B
3	84,03	COLOUR	C
4	83,04	FOLLOW	D
5	82,05	FACTORS	E
6	81,06	SQUARE	F
7	80,07	CREATOR	G
8	79,08	NATURE	H
9	78,09	AMBROSIA	I
10	77,10	MATTER	J
11	76,11	MINUS	K
12	75,12	SERIES	L
13	74,13	PAIRING	M
14	73,14	FORMAT	N
15	72,15	ORIGIN	O
16	71,16	SPHERE	P
17	70,17	STOP	Q
18	69,18	ZOOM	R
19	68,19	JOINT	S

Sr.no	Partition	First Formulation	Second Formulation
20	67,20	WATER	T
21	66,21	SINGLE	U
22	65,22	CENTER	V
23	64,23	ZERO	W
24	63,24	LIMIT	X
25	62,25	COSMIC	Y
26	61,26	CHURCH	Z
27	60,27	FOUR	HALF
28	59,28	SOLID	AIR
29	58,29	TWO	BLACK
30	57,30	HEIGHT	BIBLE
31	56,31	LIGHT	CUBE
32	55,32	SKY	LIFE
33	54,33	SUN	SEED
34	53,34	AXIS	ONE
35	52,35	EARTH	EYE
36	51,36	FULL	REAL
37	50,37	VOID	SEAL
38	49,38	AXES	ION
39	48,39	POLE	ANGLE
40	47,40	MONAD	LINE
41	46,41	LOGIC	AFFINE
42	45,42	RANGE	NEW
43	44,43	SPACE	FRAME

The following is one tabulation of words-formulations with NVFs 88 to 174. It may helped to have insight into the organization of artifices formats in the light of different partitions accepting **truth** as common formulation as one part of all pairings.

Sr.no	Partition	First Formulation	Second Formulation
1	1+87	A	TRUTH
2	2+87	B	TRUTH
3	3+87	C	TRUTH
4	4+87	D	TRUTH
5	5+87	E	TRUTH
6	6+87	F	TRUTH
7	7+87	G	TRUTH
8	8+87	H	TRUTH
9	9+87	I	TRUTH
10	10+87	J	TRUTH
11	11+87	K	TRUTH
12	12+87	L	TRUTH
13	13+87	M	TRUTH
14	14+87	N	TRUTH
15	15+87	O	TRUTH
16	16+87	P	TRUTH
17	17+87	Q	TRUTH
18	18+87	R	TRUTH
19	19+87	S	TRUTH
20	20+87	T	TRUTH
21	21+87	U	TRUTH
22	22+87	V	TRUTH
23	23+87	W	TRUTH
24	24+87	X	TRUTH
25	25+87	Y	TRUTH
26	26+87	Z	TRUTH
27	27+87	HALF	TRUTH
28	28+87	AIR	TRUTH
29	29+87	BLACK	TRUTH
30	30+87	BIBLE	TRUTH
31	31+87	CUBE	TRUTH
32	32+87	LIFE	TRUTH
33	33+87	SEED	TRUTH
34	34+87	ONE	TRUTH
35	35+87	EYE	TRUTH
36	36+87	REAL	TRUTH
37	37+87	SEAL	TRUTH
38	38+87	ION	TRUTH

39	39+87	ANGLE	TRUTH
40	40+87	LINE	TRUTH
41	41+87	AFFINE	TRUTH
42	42+87	NEW	TRUTH
43	43+87	FRAME	TRUTH
44	44+87	SPACE	TRUTH
45	45+87	RANGE	TRUTH
46	46+87	LOGIC	TRUTH
47	47+87	MONAD	TRUTH
48	48+87	POLE	TRUTH
49	49+87	AXES	TRUTH
50	50+87	VOID	TRUTH
51	51+87	FULL	TRUTH
52	52+87	EARTH	TRUTH
53	53+87	AXIS	TRUTH
54	54+87	SUN	TRUTH
55	55+87	SKY	TRUTH
56	56+87	LIGHT	TRUTH
57	57+87	HEIGHT	TRUTH
58	58+87	TWO	TRUTH
59	59+87	SOLID	TRUTH
60	60+87	FOUR	TRUTH
61	61+87	CHURCH	TRUTH
62	62+87	COSMIC	TRUTH
63	63+87	LIMIT	TRUTH
64	64+87	ZERO	TRUTH
65	65+87	CENTER	TRUTH
66	66+87	SINGLE	TRUTH
67	67+87	WATER	TRUTH
68	68+87	JOINT	TRUTH
69	69+87	ZOOM	TRUTH
70	70+87	STOP	TRUTH
71	71+87	SPHERE	TRUTH
72	72+87	ORIGIN	TRUTH
73	73+87	FORMAT	TRUTH
74	74+87	PAIRING	TRUTH
75	75+87	SERIES	TRUTH
76	76+87	MINUS	TRUTH
77	77+87	MATTER	TRUTH
78	78+87	AMBROSIA	TRUTH

79	79+87	NATURE	TRUTH
80	80+87	CREATOR	TRUTH
81	81+87	SQUARE	TRUTH
82	82+87	FACTORS	TRUTH
83	83+87	FOLLOW	TRUTH
84	84+87	COLOUR	TRUTH
85	85+87	CREATION	TRUTH
86	86+87	PARAGRAPH	TRUTH
87	87+87	TRUTH	TRUTH

The following is one tabulation of words-formulations with NVFs 88 to 174. It may helped to have insight into the organization of artifices formats in the light of different partitions accepting **truth** as common formulation as one part of all pairings.

NVF	FORMULATION	FIRST PART	SECOND PART
88	VOLUME	A	TRUTH
89	UNITY	B	TRUTH
90	ARTIFICES	C	TRUTH
91	MIRROR	D	TRUTH
92	REVERSE	E	TRUTH
93	ENTITY	F	TRUTH
94	TRI-MONAD	G	TRUTH
95	RENEWING	H	TRUTH
96	KNOWLEDGE	I	TRUTH
97	PILGRIMAGE	J	TRUTH
98	TRANSCEND	K	TRUTH
99	THOUGHT	L	TRUTH
100	DISCIPLINE	M	TRUTH
101	INTERVAL	N	TRUTH
102	TWO-SPACE	O	TRUTH
103	COUNTING	P	TRUTH
104	FOUR-SPACE	Q	TRUTH
105	DEFINITION	R	TRUTH
106	INFINITY	S	TRUTH
107	REFECTION	T	TRUTH
108	GEOMETRY	U	TRUTH
109	FIRST SEAL	V	TRUTH
110	MEDITATION/SUNLIGHT	W	TRUTH

NVF	FORMULATION	FIRST PART	SECOND PART
111	PERFECTION/TETRA-MONAD	X	TRUTH
112	MATHEMATICS	Y	TRUTH
113	UNIVERSE	Z	TRUTH
114	IMPULSES/RECITATIONS	HALF	TRUTH
115	DIMENSIONAL/INTELLIGENCE	AIR	TRUTH
116	GEOMETRIES/FOUR FOLDS	BLACK	TRUTH
117	MANIFESTING/STATEMENT	BIBLE	TRUTH
118	WONDERFUL	CUBE	TRUTH
119	FOUNDATION	LIFE	TRUTH
120	WHOLENESS	SEED	TRUTH
121	UNIFICATION/UNIVERSAL	ONE	TRUTH
122	FREQUENCIES	EYE	TRUTH
123	SYNTHETIC/CIRCUMFERENCE	REAL	TRUTH
124	TECHNOLOGY	SEAL	TRUTH
125	PROCESSING/SPIRITUAL	ION	TRUTH
126	TRUTHFUL	ANGLE	TRUTH
127	VERTICALLY	LINE	TRUTH
128	UNCOUNTABLE	AFFINE	TRUTH
129	HALF DIMENSION	NEW	TRUTH
130	CONTINUUM	FRAME	TRUTH
131	HYPER SOLID/PROFESSOR	SPACE	TRUTH
132	META PHYSICAL	RANGE	TRUTH
133	PRINT OUT	LOGIC	TRUTH
134	INTELLECTUAL/MANUSCRIPT	MONAD	TRUTH
135	INTENSITY	POLE	TRUTH
136	CONSECUTIVE	AXES	TRUTH
137	CHARACTERISTIC	VOID	TRUTH
138	SYMMETRY/SYNTHESIS	FULL	TRUTH
139	DISCIPLINARY	EARTH	TRUTH
140	CLASSIFICATION	AXIS	TRUTH
141	ORIENTATION	SUN	TRUTH
142	INTERLINKING	SKY	TRUTH
143	PHILOSOPHY	LIGHT	TRUTH
144	FORMULATION/CONCLUSIONS	HEIGHT	TRUTH
145	STRUCTURE	TWO	TRUTH
146	MANIFESTATION	SOLID	TRUTH
147	COMPUTATION	FOUR	TRUTH
148	SCRIPTURES	CHURCH	TRUTH

NVF	FORMULATION	FIRST PART	SECOND PART
149	INTER CONNECTED	COSMIC	TRUTH
150	CONTINUITY/UNIFORMLY	LIMIT	TRUTH
151	CONTINUOUS/PHYSIOLOGY	ZERO	TRUTH
152	INTERNATIONAL	CENTER	TRUTH
153	AUTOMATICALLY	SINGLE	TRUTH
154	COMPREHENSION	WATER	TRUTH
155	CONSEQUENTIAL	JOINT	TRUTH
156	PARTICULARLY	ZOOM	TRUTH
157	DIMENSIONLESS	STOP	TRUTH
158	TWO DISCIPLINE	SPHERE	TRUTH
159	CLASSIFICATIONS	ORIGIN	TRUTH
160	SEQUENTIALLY/CONSTITUEN T	FORMAT	TRUTH
161	VOLUMINOUS	PAIRING	TRUTH
162	UNIVERSITY	SERIES	TRUTH
163	SERIOUSNESS	MINUS	TRUTH
164	COMPLIMENTARY	MATTER	TRUTH
165	SUCCESSFULLY	AMBROSIA	TRUTH
166	PRESUMPTION	NATURE	TRUTH
167	IN COMPREHENSIBLE	CREATOR	TRUTH
168	EXPLORATIONS	SQUARE	TRUTH
169	SIMULTANEOUS	FACTORS	TRUTH
170	ILLUSTRATION/TRANSMISSIO N	FOLLOW	TRUTH
171	EXHAUSTIVELY	COLOUR	TRUTH
172	APPROPRIATELY	CREATION	TRUTH
173	COMPREHENSIONS	PARAGRAP H	TRUTH
174	MULTIPLICATION	TRUTH	TRUTH

NVF (MANIFEST)=NVF (TRUTH

Manifest and **Truth** are formulations of common Number Value Formats.

Pairing Discipline of spatial order pairs two as

(Two, Two)

as of NVF (Two, Two)=156

=NVF (Four Folds)

=NVF (Four Domain)

Within **Four Domain** are sustained all the First to Four Domain.

In other words, the measuring rod of 4-space is synthesis of representative regular bodies of 1,2,3 and 4 space, namely, interval, square, cube and hyper cube

The spatial order of 4 space manifests this synthesis as four fold manifestation as A dimensional, B boundary, C domain and D hyper (origin).

One may tabulate their NVFs to reach at their manifested messages as. .

NVF (A DIMENSION)=103=NVF (COUNTING),

NVF (B BOUNDARY)=102=NVF (TWO SPACE),

NVF (C DOMAIN)=59=NVF (SOLID),

NVF (D ORIGIN)=76=NVF (MINUS),=NVF (ADDITION),

NVF (DIMENSION)=102=NVF (TWO SPACE),

NVF (BOUNDARY)=100=NVF (DISCIPLINE)

=NVF (HUMAN FRAME)

=NVF (SLEEP FRAME)

=NVF (A DOMAIN FRAME)

NVF (HYPER, ORIGIN=144=NVF (SPACE DISCIPLINE)

One may chase the manifestation phenomena in reference to the cube and appreciate manifestations of different folds with 1 space in the role of dimension fold, 2 space in the role of boundary fold, 3 space in the role of domain fold and 4 space in the role of origin fold.

NVF(DIMENSION)=102=NVF(TWO SPACE) and NVF(INTERVAL)=101=NVF(ULTIMATE)=NVF(DIVISION)

When simultaneously worked, makes interval as a formulation of format of di-monad line.

Format itself is a formulation, which as well avails di-monad line as on format. The di-monad line is a synthesis of a pair of lines, as of two parts of di-monad.

Di-monad is organised in opposite orientations as reflection pair of images of each other in reference to the mirror embedded at the joint of the parts of the di-monad.

This is an expression along the artifices of numbers with joint (POINT) as '0' and ends on either side as '1' as an expression '101', with one part from one end up-till middle (JOINT) as '10' and the second part form middle joint up-till other and as "01"

Further, NVF (TEN)+NVF (ONE)=39+34=73=NVF (FORMAT) will help us comprehend and to appreciate the manifested format of this organization of formulations of artifices and formats where in every aspect is inter-related, inter-connected, coordinated & correlated with Reality of Existence phenomena within frames in a space fulfilled with light from orb of the Sun.

As such, sadkhas as students knowing counting

NVF (STUDENT)=103=NVF (COUNTING) may initiate themselves by availing formats of artifices of numbers for the transcending mind being transcending line as

NVF (MIND)=40=NVF (LINE)

To transcend is to reach at the bed of Truth/Manifest bed as

NVF(TRANSCEND)=98=87+11

=NVF(TRUTH BED)

=NVF(MANIFEST BED)

From "Manifest" to "Transcend", there is a continuity of steps, as much as that the format of artifice '87' takes from artifice 7 to artifice 8 while the format of artifices "98" takes from artifices 8 to artifices 9.

With this, the achievement is full and complete: the unity state of consciousness flourishing and blossoming blissfully as ambrosia of the Ultimate Brahmin as Nav-Brahmin/9-Space, as

NVF(UNITY) =89, brings back from 9 space (NAV BRAHM) to 8 space (Asht-Prakriti) along the range taking from artifice 9 to artifice 8

NVF(AMBROSIA) =78, brings back from 8 space (Asht-Prakriti) to 7 space (SAPT-RISHI-LOK) along the range taking from artifice 8 to artifice 7 where is sustained is field of unity state of consciousness to flourish as existence phenomena with in human frame as boundary discipline for a domain frame with in human frame.

The consciousness field sustained by Sapt-rishi lok is of Seven states, namely A Waking, B Dreaming, C Deep Sleep, D transcendental, E Cosmic, F God and G Unity state of consciousness.

Sadkhas with an intensified urge to reach unity state of consciousness shall

(a) sit comfortably and,

(b) permit the mind to transcend

- to glimpse the transcendental world,

as transcendental phenomena,

and have, self validation of the way and path,

of transcendental glimpse of vedic mathematics.

Between two sitting of trans,

The Sadhkas, shall,

(a) go for the intellectual exercise of to be through the scriptures,

(b) and also share the experiential bliss of transcendental world with fellow Sadkhas,

(c) and well as to be sure about the perfection of intelligence gained through transcendence and through intellectual exercise.

Sadhkas shall also take it a blissful exercise to experience during trans chase of the higher spaces reality.

■

6
EXISTENCE OF HIGHER SPACES

The intellectual chase of these phenomena of existence of higher spaces may be on the lines of lessons 7 & 8 of "Learn and Teach Vedic Mathematics" book of the author. These lessons are being re-produced here for convenient reference.

SOME SPECIFIC FEATURES OF SPACE BOOK

1. **Space book**

   NVF(SPACE BOOK)=87

   =NVF(SPACE PLAN)

   =NVF(SPACE FRAME)

   =NVF(TRUTH)

2. **Fore-word**

   NVF(FORE-WORD)=104

   =NVF(FOUR SPACE)

3. **Pre-face**

   NVF(PRE-FACE)=54

   =NVF(SUN)

4. **Subject**

NVF(SUBJECT)=80

=NVF(CREATOR)

5. Paper

NVF(PAPER)=56

=NVF(LIGHT)

6. Pen

NVF(PEN)=35

=NVF(EYE)

7. Ink

NVF(INK)=34

=NVF(CONE)

=NVF(DARK)

=NVF(DUDE)

8. Author

NVF(AUTHOR)=83

=NVF(FOLLOW)

=NVF(BLACK SUN)

=NVF(DARK LORD)

=NVF(ONE LORD)

9. Contents

NVF(CONTENTS)=110

=NVF(MIRRORS)

10. Sections

NVF(SECTIONS)=104

=NVF(EXISTENCE)

=FOUR SPACE

11. Chapters

NVF(CHAPTERS)=90

=NVF(ARTIFICES)

=NVF(SPHERES)

=NVF(NEW TREE)

12. Lessons

NVF(LESSONS)=103

=NVF(COUNTING)

=NVF(STUDENT)

=NVF(HEAVEN TREE)

13. Object

NVF(OBJECT)=55

=NVF(HEAVEN)

=NVF(SKY)

14. Theme

NVF(THEME)=51

=NVF(FULL)

15. Aim

NVF(AIM)=23

=NVF(END)

16. Goal

NVF(GOAL)=38

=NVF(ION)

=NVF(FIRE)

17. Conclusions

NVF(CONCLUSIONS)=144

=NVF(SPACE DISCIPLINE)

=NVF(HYPER ORIGIN)

18. Virtue

NVF(VIRTUE)=95

=NVF(RENEWING)

19. Bliss

NVF(BLISS)=61

=NVF(CHURCH)

20. Fruit

NVF(FRUIT)=74=

NVF(PAIRING)

21. Meditate

NVF(MEDITATE)=77

=NVF(CHRIST)

22. Static

NVF(STATIC)=72

=NVF(ORIGIN)

23. Moving

NVF(MOVING)=80

=NVF(CREATOR)

24. Beginning

NVF(BEGINNING)=84

=NVF(COLOUR)

=NVF(GOD TWO)

25. Gods

NVF(GODS)=45

=NVF(RANGE)

26. God One, Two, Three, Four, Five, Six

NVF(GOD ONE)=60=NVF(FOUR)

NVF(GOD TWO)=84=NVF(COLOUR)

NVF(GOD THREE)=92=NVF(REVERSE)

NVF(GOD FOUR)=86=NVF(NEW SPACE)

NVF(GOD FIVE)=68=NVF(JOINT)

NVF(GOD SIX)=78=NVF(AMBROSIA)

FOCUS UPON SOME FEATURES

The space plan truth features are reflected in the groups of words-formulations accepting common number value formats. Here below are being focused some of these sub-sets of formulations :

1. NVF(SPHERE)+NVF(SPHERE)

   =71+71=14 =102+40

   =NVF(TWO SPACE LINE) =NVF(STRAIGHT LINE)

2. NVF(ORIGIN)+NVF(ORIGIN)

   =72+72=144=100+44

   =NVF(SPACE DISCIPLINE)=NVF(INTERVAL FRAME)

3. NVF(MONAD)+NVF(MONAD)

   =47+47=94=NVF(TRI-MONAD)

4. NVF(CONE)+NVF(CONE)=37+37=74

   =NVF(ENGLISH)=NVF(PAIRING)

   =NVF(POINT) =NVF(FRUIT)

5. NVF(BLACK)+NVF(BIBLE)=29+30=59

   =NVF(LINEAR)=NVF(DOUBLE)=NVF(SOLID)

6. NVF(MIND)+NVF(SEED)=40+33=73

   =NVF(FORMAT)=NVF(SOUND)=NVF(PULSE)

7. NVF(MIND)+NVF(SEAL)=40+37=77

   =NVF(MATTER)=NVF(MEDIATE)=NVF(CHRIST)

8. NVF(TRANSCENDENTAL)=NVF(SELF-REFERRAL)

TRANSITION LEAD

The space plan is of such transcendental features for which range, the sadkhas shall go in trans time and again and glimpse the inner folds of the transcendental world and to perfect ones intelligence to chase their transcendental values.

VEDIC MATHEMATICS DECODES SPACE BOOK.

■

7
PLAYING RAMNUJAN

The students may refer to the book (VEDIC MATHEMATICS DECODES SPACE BOOK) for perfection of their intelligence by permitting the mind to transcend as well as by playing Ramanujan with cardinal no 26 availing the following four sets of 26 artifices:

1. Counts: (26)

   1,2,3,4,5,6,7,8,9,10,11,12,13,14,15,16,17,18,19 ,20,21,22,23,24,25,26

2. Primes: (26) including 1

   1,2,3,5,7,11,13,17,19,23,29,31,37,41,43, 47,53,59,61,67,71,73,79,83,89,97

3. Primes: (26)

   2,3,5,7,11,13,17,19,23,29,31,37,41,43,47, 53,59,61,67,71,73,79,83,89,97,101

4. Odd Primes: (26)

   3,5,7,11,13,17,19,23,29,31,37,41,43,47, 53,59,61,67,71,73,79,83,89,97,101,103

   Play Ramanujan game of cardinal number 26 with

word formulation **one** in terms of above four sets of artifices.

Hint 1: The value/number value format of this formulation (one) for the first set is o=15+n=14+e=5 total 15+14+5=34

Hint 2: The value/number value format of this formulation (one) for the second set is o=43+n=41+e=7 total 43+41+7=91

Hint 3: The value/number value format of this formulation (one) for the third set is o=47+n=43+e=11 total 47+43+11=101

Hint 4: The value/number value format of this formulation (one) for the fourth set is o=53+n=47+e=13 total 53+47+13=113

Hint 5: NVF (ONE)=34 for the first set:

NVF (MIRROR)=91 for the first set:

NVF (INTERVAL)=(101) for the first set:

NVF (UNIVERSE)=(113) for the first set:

Hint 6: The formulation one as one, mirror, interval and universe are the manifested values of same formulation within different folds.

Like formulation **one,** each word formulation has distinct four folds manifested values within different folds whose chase is to make a complete chase of the space plan in its each feature, series of features, strings of series of features, sequences of strings of series of features required for zooming and netting of ultimate void (151) as full discipline (151) to provide ultimate void bed (162) for TRINITY OF GODS manifesting as COSMIC DISCIPLINE as INTELLIGENCE MONAD.

A few Ramanujam Game intellectual sittings

Workout, orally, the number value formats for the following **words-formulations** for all the **six situations** of alphabet letters accepting NVF for them, **firstly**, as artifices 1 to 26, **secondly** as first 26 primes (including one), **thirdly** as first 26 primes (excluding one), **fourthly** for first 26 odd primes, and **fifthly** for first 26 primes beginning with 5 as the first (transcendental) prime and **sixthly** for the first 26 beginning with 7 as the first (unity) prime:

[for the facility of convenient reference, the primes, (including one), are being enlisted as 1, 2, 3, 5, 7, 11, 13, 17, 19, 23, 29, 31, 37, 41, 43, 47, 51, 53, 59, 61, 67, 71, 73, 79, 83, 89, 97, 101, 103, 107, 109)

For further convenience, following table is being drawn

Sr. no	Alphabet letter	NVF1st	NVF2nd	NVF 3rd	NVF4th	NVF5thNVF
1	A	1	1	2	3	5
2	B	2	2	3	5	7
3	C	3	3	5	7	11
4	D	4	5	7	11	13
5	E	5	7	11	13	17
6	F	6	11	13	17	19
7	G	7	13	17	19	23
8	H	8	17	19	23	29

Sr. no	Alphabet letter	NVF 1st	NVF 2nd	NVF 3rd	NVF 4th	NVF 5th NVF
9	I	9	19	23	29	31
10	J	10	23	29	31	37
11	K	11	29	31	37	41
12	L	12	31	37	41	43
13	M	13	37	41	43	47
14	N	14	41	43	47	53
15	O	15	43	47	53	59
16	P	16	47	53	59	61
17	Q	17	53	59	61	67
18	R	18	59	61	67	71
19	S	19	61	67	71	73
20	T	20	67	71	73	79
21	U	21	71	73	79	83
22	V	22	73	79	83	89
23	W	23	79	83	89	97
24	X	24	83	89	97	101
25	Y	25	89	97	101	103
26	Z	26	97	101	103	107

Help may be taken from the above table for NVFs of individual letters and the NVFs of the following words formulations may be computed for the given five situations, and the values of NVFs for second to fifth situations to be compared with the words formulations of parallel values for the first situation. Here below, first of all is being tabulated such NVFs for the word formulation 'ONE'.

Sr. Word 1st NVF 2nd

no Formulation

NVF 3rd

NVF 4th

NVF 5th

NVF

1 ONE 15+14+5 43+41+7 47+43+11
53+47+13 59+53+17

2 ONE 34 91 101 113 129

3 ONE ONE MIRROR INTERVAL
UNIVERSE HALF DIMENSION

Here is the list of the words formulations which be taken up as exercises of playing Ramanujam, to reach at the transformations for the words formulations like 'one' being of five formats as 'one, mirror, interval, universe and half dimension'.

List of words formulations for Ramanujam play exercises

'Two, three, four, five, six, seven, eight and nine'.

Like that a big range of series of words-formulations can be taken up for exercises of Ramanujam play as steps for perfection of intelligence. Here below is being taken up the exercise formulation, namely, 'TWO'. Its chain of five formulations would be as follows:

Sr.

No Word 1st

NVF 2nd

NVF 3rd

NVF 4th NVF 5th NVF

1 TWO 20+23+15 67+79+43
71+83+47 73+89+53 79+97+59

2 TWO 58 189 201 215 235

3 TWO TWO UNITY DISCIPLINE / ULTIMATE
VOLUME ULTIMATE DISCIPLINE
DISCIPLINE INTELLIGENCE ULTIMATE ONE
DISCIPLINE

It would be blissful exercise to compare the five situations range of one with five situations range of ‘two’. While the situations range of ‘one’ comes to be ‘one, mirror, interval, universe and half dimension’, the same for ‘two comes to be ‘two, ultimate volume, ultimate discipline, discipline intelligence, ultimate one discipline’.

Like that the situations ranges for ‘three’, four, five and so on may be workout and can be compared inter see and playing like that one may expect to be ‘Ramanujam’.

For further intellectual chase

The book ‘Vedic mathematics decodes SPACE BOOK’ of the Author published by lotus press Delhi, may be referred for further insight into the intrinsic values of the number value formats of words formulations accepting pairing parallel to pairing of artifices 1 to 26, carrying along formats of 26 basic elements, as 26 primes and also 26 entities constituting synthetic boundary for the full range of cube to cuboid / cube end.

■

SECTION - 5

TRANSCENDENTAL NATURE OF VEDIC MATHEMATICS

1
AMBROSIA OF BLISS, IMPULSES OFCONSCIOUSNESS AND FREQUENCES OF NAD AND JYOTI

AMBROSIA OF BLISS

Transcending mind,

(a) glimpses the transcendental world,

(b) and is full filled with ambrosia of bliss,

which when shared by *Sadhkas* with fellow *Sadhkas*,

(a) manifests,

(b) as impulses of consciousness.

Sadhkas having intensified urge to know and chase this transcendental phenomena of manifestation of ambrosia of bliss on its sharing by *Sadhkas* manifesting as impulses of consciousness,

- may go for its self-validation,
- by glimpsing the transcendental world,
- and by sharing the bliss with the fellow

Sadhkas.

IMPULSES OF CONSCIOUSNESS

The transcendental phenomena,

- sharing, of experiential bliss manifesting as impulses of consciousness,

is well preserved in the scriptures.

These impulses of consciousness are manifested as the frequencies of *Nad* and *Jyoti.*

This is transcendental phenomena of sharing of experiential bliss.

FREQUENCIES OF NAD AND JYOTI.

The frequencies,

(a) of *Nad,*

(b) and *Jyoti,*

manifest,

(a) as the artifices formats,

(b) availed for organization of knowledge,

(i) as *Sutras,*

(ii) Shalokas,

(iii) Mantras,

(iv) and *Satotrams,*

and these frequencies are,

(a) preserved in the scriptures,

(b) and the same are availed by the *Sadhka,*

(i) as projected thoughts,

(ii) for manifesting,

- artifice formats,
- for the *Ati-Vahkas,*
- for carrying the transcending mind,
- as well as for carrying the Being,

(c) through *SushmanaNaadi,*

(i) uptil *Brahamrander,*

(ii) and from there through rays of the Sun uptil core of the Sun,

(iii) and for onward pilgrimage Being carried by *Ati-Vahkas* of *Sapat-Rishi-Lok,*

to the *Brahaman* domain.

SUTRAS, SHALOKAS, MANTRAS AND SATROTAMS

SUTRAS

Sutras are the frequencies,

(a) of *Nad,*

(b) and *Jyoti,*

manifestation of,

(a) impulses of consciousness,

(b) of *Sadhkas* having glimpsed the transcendental world.

Literally, *Sutra,* means thread, the stitching thread of three folds of the *Ganas* and it is removal of these stitching threads of *Ganas* which manifests the Discipline of *Ganita* (Mathematics).

■

2

TRANSCENDENTAL FEATURES OF VEDIC MATHEMATICS

1. Vedic systems organized artifices of numbers on various place value format. Of these, one of various applied values is ten-place value format.
2. Here, unless the context is otherwise, the numbers expressions format taken being availed is ten-place value format.
3. Being ten place value systems it makes available nine positive numerals as counts one to nine. In addition, replace value format, provides '0' numeral.
4. As such, the basics of the arithmetic operations need be approached first in reference to these ten numerals namely '0,1,2,3,4,5,6,7,8,9'.
5. This in a way to have arithmetic operations of these ten numerals with themselves.
6. For basic arithmetic operations are 'addition, minus, multiplication and division'.
7. These operations have been provided special notational symbols, viz., + for addition, - for minus, × for multiplication, and for division.

8. The symbol for addition operation (+) is a symbol of a pair of axes, symbolsing and indicating that the geometric format for the operation is as would be manifesting attainment from a given single axes leading to the pair of axes.
9. The symbol for minus operation (-), is a symbol of a single axis, which in reference to the about symbol (+) shall be a reversal from a pair of axes to a single axis, as a reversal of the addition operation, and hence designating as minus.
10. The symbol for multiplication (×) as comparison to the addition (+) is to attain domain placement, for axes, in place of their boundary placement (of plane/surface). This is a shift from boundary to domain; linear order to spatial order.
11. Multiplication as repeated addition, and the symbolic notations in terms of axes indicating a shift from a boundary to domain deserve to be chase to appreciate the unison of the processes as well as their distinguishing features because of which these (addition and multiplication) emerge to be distinct arithmetic operations, as well, as that of arithmetic operations are ultimately of same origin of artifices of numbers accepting counts as their values manifesting as counting with the rule of 'one more than before' and thereby the whole range of values getting sequenced as a single range.
12. Multiplication symbol is segregation around the axis as a pair of parts, with a point to point correspondence along the pair of parts of the second axis. This operation, as such is a reversal of the form and format which makes it in reference to multiplication, as is minus related to addition.

3

ONE OUTLINE OF SOME STEPS

Here is being drawn one outline of some steps which may be of help for the students, parents, teachers and seniors:

1. While sitting in a room, one gets condition as if the universe is just 3-space.
2. For melting of the 3-space mental block, one is to come out of the room and to consciously expose oneself to the reality under the Sun.
3. For perfection of intelligence, teachers shall take gentle initiatives for young minds to melt the emerging mental blocks. These initiatives, amongst other features, may also include a planned programme for sadkhas to have living for a day and a night under different environments then that of a living room of a home and that of a classroom in a school.
4. Amongst different environments, living in an Ayurvedic garden may be one such choice.
5. Within Ayurvedic garden, living for a day and a night on a big tree would be further specific choice.

6. Living on a boat in a lake for a day and night is one such another exposure.
7. Living alone in a jungle for a day and a night without food may be another choiced experience.
8. Chasing moon during night remaining awake is going to be of fruitful initiative watching the way small creature lives his life during a day and a night and its comparison with one's own is going to be blissful.
9. Living a day and night with eyes closed is another domain of experiential bliss.
10. Silence has its own message.
11. Reciting a scripture for a day and a night has its own reward.
12. Thinking for a day and night about oneself is of far reaching rewards.
13. Permitting the mind to transcend is blissful.
14. Transcending mind chasing itself is self-referral.
15. Glimpsing inner folds and inner most fold of the transcendental world is enlightenment.

■

4

ORGANISATION FORMAT OF '1 TO 9'

ORGANIZATION FORMAT OF "1" AS 1 SPACE

1. Sadhkas shall sit comfortably and avail organization format of "1" as 1 space to perfect one's intelligence along the **first** step of the measuring rod as of the format of representative regular body of **1** space within transcendental world.
2. It would be a blissful exercise to enlist One's transcendental experiences of glimpsing inner folds along the organization format of "1" as 1 space and also as 1-manifestation layer (-2,-1,0,1) with –2 space in the role of dimension, -1 space as boundary, 0 space as domain and 1 space as origin..
3. One shall share one's experiences and intellectual comprehensions of organization of "1" with fellow Sadhkas.
4. Sadhka shall continuously up to date oneself about one's comprehensions of organization format of "1"

after experiencing transcendence along the comprehended features of organization format of "1".

5. It shall be very fruitful intellectual exercise to enlist formulations of different attributes of "1" and its organization format, and also to chase the geometric manifestations of such attributes / features / properties / characteristics of artifice "1" like 1-Space, linear order, Unit, Sole etc. etc.

ORGANIZATION FORMAT OF "2" AS 2 SPACE

1. Sadhkas shall sit comfortably and avail organization format of "2" as 2 space to perfect one's intelligence along the **second** step of the measuring rod as of the format of representative regular body of **2** space within transcendental world.
2. It would be a blissful exercise to enlist One's transcendental experiences of glimpsing inner folds of the organization format of "2" as 2 space and also as 2-manifestation layer (-1,0,1,2) with –1 space in the role of dimension, 0 space as boundary, 1 space as domain and 2 space as origin..
3. One shall share one's experiences and intellectual comprehensions of organization of "2" with fellow Sadhkas.
4. Sadhka shall continuously up to date oneself about one's comprehensions of organization format of "2" after experiencing transcendence along the comprehended features of organization format of "2".

5. It shall be very fruitful intellectual exercise to enlist formulations of different attributes of "2" and its organization format, and also to chase the geometric manifestations of such attributes/ features/properties/characteristics of artifice "2" like 2-Space, spatial order, pair, di-monad etc. etc.

ORGANIZATION FORMAT OF "3" AS 3 SPACE

1. Sadhkas shall sit comfortably and avail organization format of "3" as 3 space to perfect one's intelligence along the **third** step of the measuring rod as of the format of representative regular body of **3** space within transcendental world.
2. It would be a blissful exercise to enlist One's transcendental experiences of glimpsing inner folds of the organization format of "3" as 3 space and also as 3-manifestation layer (0,1,2,3) with 0 space in the role of dimension, 1 space as boundary, 2 space as domain and 3 space as origin..
3. One shall share one's experiences and intellectual comprehensions of organization of "3" with fellow Sadhkas.
4. Sadhka shall continuously up to date oneself about one's comprehensions of organization format of "3" after experiencing transcendence along the comprehended features of organization format of "3".
5. It shall be very fruitful intellectual exercise to enlist formulations of different attributes of "3"

and its organization format, and also to chase the geometric manifestations of such attributes/ features/properties/characteristics of artifice "3" like 3-Space, Solid order, Triplet, tri-monad, tri-angle etc. etc.

ORGANIZATION FORMAT OF "4" AS 4 SPACE

1. Sadhkas shall sit comfortably and avail organization format of "4" as 4 space to perfect one's intelligence along the **fourth** step of the measuring rod as of the format of representative regular body of **4** space within transcendental world.
2. It would be a blissful exercise to enlist One's transcendental experiences of glimpsing inner folds of the organization format of "4" as 4 space and also as 4-manifestation layer (1,2,3,4) with 1 space in the role of dimension, 2 space as boundary, 3 space as domain and 4 space as origin.
3. One shall share one's experiences and intellectual comprehensions of organization of "4" with fellow Sadhkas.
4. Sadhka shall continuously up to date oneself about one's comprehensions of organization format of "4" after experiencing transcendence along the comprehended features of organization format of "4".
5. It shall be very fruitful intellectual exercise to enlist formulations of different attributes of "4" and its organization format, and also to chase the geometric manifestations of such attributes/

features/properties /characteristics of artifice "4" like 4-Space, Hyper Solid order-4, etc. etc.

ORGANIZATION FORMAT OF "5" AS 5 SPACE

1. Sadhkas shall sit comfortably and avail organization format of "5" as 5 space to perfect one's intelligence along the **fifth** step of the measuring rod as of the format of representative regular body of **5** space within transcendental world.
2. It would be a blissful exercise to enlist One's transcendental experiences of glimpsing inner folds of the organization format of "5" as 5 space and also as 5-manifestation layer (2,3,4,5) with 2 space in the role of dimension, 3 space as boundary, 4 space as domain and 5 space as origin.
3. One shall share one's experiences and intellectual comprehensions of organization of "5" with fellow Sadhkas.
4. Sadhka shall continuously up to date oneself about one's comprehensions of organization format of "5" after experiencing transcendence along the comprehended features of organization format of "5".
5. It shall be very fruitful intellectual exercise to enlist formulations of different attributes of "5" and its organization format, and also to chase the geometric manifestations of such attributes/ features/properties/characteristics of artifice "5" like 5-Space, Hyper Solid order-5, etc. etc.

ORGANIZATION FORMAT OF "6" AS 6 SPACE

1. Sadhkas shall sit comfortably and avail organization format of "6" as 6 space to perfect one's intelligence along the **Sixth** step of the measuring rod as of the format of representative regular body of **6** space within transcendental world. Here 6 space within 5 space is in the role of origin of 5 space. Further 6 space being of hyper dimensional order as much as that 4 space place the role of dimension of 6 space, as such second creation as of inner fold gets initiative and as such there emerges transitions from macro level existence of 6 space as domain to micro level existence of 4 space as dimension. This shift and transition takes from domain to dimension. As such there is a shift to the manifestation layer with 6 space as origin fold with three space as dimension fold. This transcendental feature of transitions and shift to manifestation layer as micro state bodies is the phenomena of space plan which deserves to be chased as happening as a self-referral transcendental phenomena. With it the organisation format of 6 space as chased in terms of Sathapatya measuring rod takes to chasing of the inner folds of the transcendental world on format of 6 space as manifestation layer (3,4,5,6) as of 3 space as dimension fold, 4 space as boundary fold, 5 space as domain and 6 space as origin fold.
2. It would be a blissful exercise to enlist One's transcendental experiences of glimpsing inner

folds of the organization format of "6" as 6 space and also as 6-manifestation layer (3,4,5,6) with 3 space in the role of dimension, 4 space as boundary, 5 space as domain and 6 space as origin.

3. One shall share one's experiences and intellectual comprehensions of organization of "6" with fellow Sadhkas.
4. Sadhka shall continuously up to date oneself about one's comprehensions of organization format of "6" after experiencing transcendence along the comprehended features of organization format of "6".
5. It shall be very fruitful intellectual exercise to enlist formulations of different attributes of "6" and its organization format, and also to chase the geometric manifestations of such attributes/features/properties/characteristics of artifice "6" like 6-Space, Hyper Solid order-6, etc. etc.

ORGANIZATION FORMAT OF "7" AS 7 SPACE

1. Sadhkas shall sit comfortably and avail organization format of "7" as 7 space to perfect one's intelligence.

   Here 7 space within 5 space is as the 5 space playing the role of dimension of 7 space. This is the transcendental feature of the transcendental phenomena where with the grace of the lord of transcendental worlds, the sadkhas straight way attain the unity state of consciousness.
2. It would be a blissful exercise to enlist One's transcendental experiences of glimpsing inner

folds of the organization format of "7" as 7 space and also as 7-manifestation layer (4,5,6,7) with 4 space in the role of dimension, 5 space as boundary, 6 space as domain and 7 space as origin.

3. One shall share one's experiences and intellectual comprehensions of organization of "7" with fellow Sadhkas.
4. Sadhka shall continuously up to date oneself about one's comprehensions of organization format of "7" after experiencing transcendence along the comprehended features of organization format of "7".
5. It shall be very fruitful intellectual exercise to enlist formulations of different attributes of "7" and its organization format, and also to chase the geometric manifestations of such attributes/ features/properties/characteristics of artifice "7" like 7-Space, Hyper Solid order-7, etc. etc.

ORGANIZATION FORMAT OF "8" AS 8 SPACE

1. Sadhkas shall sit comfortably and avail organization format of "8" as 8 space to perfect one's intelligence.

   Here 8 space within 5 space is as the 5 space straight a way fulfilling the sadkhas with the privilege potentialities of the Nature (Asht Prakrti) of the order of the transcendental world. This amounts to shifting to the manifestation layer (5,6,7,8) with 5 space in the role of dimension, 6 space in the role of boundary, 7 space in the role of domain and 8 space in the role of origin.

2. It would be a blissful exercise to enlist One's transcendental experiences of glimpsing inner folds of the organization format of "8" as 8 space and also as 8-manifestation layer (5,6,7,8) with 5 space in the role of dimension, 6 space as boundary, 7 space as domain and 8 space as origin.
3. One shall share one's experiences and intellectual comprehensions of organization of "8" with fellow Sadhkas.
4. Sadhka shall continuously up to date oneself about one's comprehensions of organization format of "8" after experiencing transcendence along the comprehended features of organization format of "8".
5. It shall be very fruitful intellectual exercise to enlist formulations of different attributes of "8" and its organization format, and also to chase the geometric manifestations of such attributes/ features/properties/characteristics of artifice "8" like 8-Space, Hyper Solid order-8, etc. etc.

ORGANIZATION FORMAT OF "9" AS 9 SPACE

1. Sadhkas shall sit comfortably and avail organization format of "9" as 9 space to perfect one's intelligence.

   Here 9 space within 5 space is as the 5 space straight a way fulfilling the sadkhas with the privilege grace of Nav-Brahm This amounts to shifting to the manifestation layer (6,7,8,9) with 6 space in the role of dimension, 7 space in the role of boundary, 8 space in the role of domain and 9 space in the role of origin.

2. It would be a blissful exercise to enlist One's transcendental experiences of glimpsing inner folds of the organization format of "9" as 9 space and also as 9-manifestation layer (6,7,8,9) with 6 space in the role of dimension, 7 space as boundary, 8 space as domain and 9 space as origin.
3. One shall share one's experiences and intellectual comprehensions of organization of "9" with fellow Sadhkas.
4. Sadhka shall continuously up to date oneself about one's comprehensions of organization format of "9" after experiencing transcendence along the comprehended features of organization format of "9".
5. It shall be very fruitful intellectual exercise to enlist formulations of different attributes of "9" and its organization format, and also to chase the geometric manifestations of such attributes/features/properties/characteristics of artifice "9" like 9-Space, Hyper Solid order-9, etc. etc.

■

5

TRANSCENDENTAL GLIMPSE OF VEDIC MATHEMATICS

CONTENTS

SECTION-1 -URGE TO KNOW

SECTION -2

ABSOLUTE FORMAT AS REFERENCE FRAME

SECTION-1 URGE TO KNOW

1.01 URGE TO KNOW

1. *Sadhkas*,
   having
   - (a) having an intensified urge,
   - (b) to know,
   - (c) and chase,
     - the transcendental glimpse of *Vedic* mathematics

   shall,
   - (a) sit comfortably and,
   - (b) permit the mind to transcend,
     - to glimpse the transcendental world,
     - as transcendental phenomena,

   and have,
   - (a) self validation of,
   - (b) the way and path,
     - of transcendental glimpse of *Vedic* mathematics.
2. Between two sittings of trans,
   the *Sadhkas*, shall,
   (a) go for the intellectual exercise of to be through the scriptures, (b) and also share the experiential bliss of transcendental world with

fellow *Sadhkas*, (c) and have self validation of experiences, as well as to be sure about the perfection of intelligence gained through transcendence and through intellectual exercise.

3. *Sadhkas* for to be sure about one's perfection of intelligence about the discipline of internal folds of the transcendental world shall enlist one's experiences and share them with fellow *Sadhkas* in the sequence those are gained and also for their validation form the scriptures. This exercise, oral or written, would be of great help, at least, at the initial stages, as the transcendence to the inner folds of the transcendental world is a very delicate exercise for the transcending mind and the transcendental world also unfolds and folds back of its own and for it the chase of the transcending mind of the transcendental phenomena of folding and unfolding of the inner folds of the transcendental world and the transcendence progress of the transcending mind, both are to be chased by the transcending mind itself.
4. The whole range of the scriptures is unfolding as the transcendental glimpse of *Vedic mathematics* and it is available with the *Sadhkas* to have it as transcendence range for satisfying the urge and full filing the transcending mind with the ambrosia of bliss of the inner most fold of the transcendental world.
5. The 'transcendental glimpse of *Vedic Mathematics*' is, the transcendental glimpse of *Vedic Mathematics* of the transcendental world.

6. This chase of the 'the transcendental glimpse of *Vedic* Mathematics' may also to mean as the *Vedic* Mathematical way and path, of glimpsing the transcendental world.

7. As such, the urge to know and chase the way and path of *Vedic* mathematics glimpsing the transcendental world is,
   - (a) the unique way and path of *Vedic* mathematics,
     - (i) approaching the transcendental world,
     - (ii) as unfolding and,
     - (iii) folding of the Reality,
   - (b) as transcendental phenomena,
     - (i) of the transcendental world,
     - (ii) with *Lord Shiv*(卐), five head Lord with three eyes in each head, as the over lord.

8. The *Sadhkas,*
   - (a) as Being within human frame,
   - (b) of one head,
   - (c) equipped with a pair of eyes,

   as such, is,
   - (a) to accept it as the start with reality,
   - (b) and it is to be the start with stage with which one is,

   to begin to go for,
   - (a) the satisfaction of the urge to know,
   - (b) and attain the order of the transcendental world of solid dimensional order of the format of idol of *Lord Shiv* (卐) as five head lord with three eyes in each head.

9. The way and path,
   (a) of chase of the transcendental glimpse of *Vedic* Mathematics of the transcendental world,
   (b) takes the *Sadhkas*,
      (i) from the start with state of the format of one head with a pair of eyes,
      (ii) to the attainable states of the format of the idol of *Lord Shiv*(卐) of five heads with three eyes in each head,
   (c) and with it,
      (i) the range of chase for the *Sadhka* gets manifested,
      with one head with pair of eyes as starting point,
      and at the end the head to get equipped with an additional eye,
      (ii) and the chase would be completed only with attainment of full chase of the transcendental glimpse of *Vedic* Mathematics uptil the inner most fold of the transcendental world.

10. *Sadhkas* experiences as are,
   (a) preserved as,
      (i) scriptures,
      (ii) traditions,
      (iii) *Maha-Mantras,*
      (iv) *Satotrams,*
      (v) *Mantras,*
      (vi) *Yantras,*
      (vii) *Tantras,*

(viii) *Jantris,*
(ix) idols,
(x) temples,

(b) and the Disciplines of,

(i) *Ganita,*
(ii) *Sankhya,*
(iii) and *Sathaptya,* in general,

(c) and *Shiva-Cult,* in its totality and applied values range in particular,

approach the transcendental world,

(i) as Reality of *Shiv-Lok,*
(ii) unfolding and folding of its own,

shall go,

(i) in trans, time and again,
(ii) and have self validation of his transcendental experiences,
  (a) by sharing the same with fellow *Sadhka,*
  (b) and by having intellectual explorations of the scriptures,
  (c) and devotional attendance to the practices of *Sadhna* practiced by other fellow *Sadhkas.*

1.02. TRANSCENDENTAL GLIMPSE OF VEDIC MATHEMATICS

11. The *Sadhkas* having an urge to know and chase the way and path of transcendental glimpse of *Vedic* Mathematics of the transcendental world as of the range with starting point as of format of one head with a pair of eyes to the attainable goal of the order of five heads with each head equipped with three eyes, shall at experiential as well as at

intellectual level have it as first self validation exercise about this as range within the above starting point and attainable point of transcendental glimpse of *Vedic* mathematics.

1.03. STARTING POINT OF THE RANGE

12. The starting point of the range is the Reality of human frame as of one head with a pair of eyes.
13. Human frame with one head as a pair of eyes is,
    - (a) the creation, of the creator,
    - (b) about whose enlightenment,

    the scriptures are,
    - (a) full of preservation of the *Shrutis*,
    - (b) as the experiential expressions,
    - (c) shared by the *Sadhkas*,
    - (d) having glimpsed the transcendental world,

    as that,
    - (a) the creator the supreme,
    - *(b) Lord Brahma* (卐),

    is the,
    - (i) four head lord,
    - (ii) with each head equipped with a pair of eyes and,

    that creator has,
    - (i) created the creations,
    - (ii) on His idol's format,
    - (iii) and of the order and discipline,
    - (iv) which permits transcendence,
        - (a) to its transcendental base of transcendental world,
        - (b) with *Lord Shiv*(卐) as over lord.
14. Scriptures are further full of preservation of the *Shruti*, the experiential expression shared by the *Sadhkas* having glimpsed the transcendental world as that:

(a) *Lord Shiv*([symbol]) is the lord of creator the supreme,

(b) and the seat of *Lord Shiv*([symbol]) is in the cavity of heart of *Lord Brahma*([symbol]).

15. Still further, scriptures are full of preservation of the *Shruti*, experiential expressions shared by the *Sadhkas* having glimpsed the transcendental world, as that:

(a) the creator the supreme, *Lord Brahma*([symbol]), meditates in his cavity of heart upon his lord, *Lord Shiv*([symbol]),

(b) and gets multiplied ten fold, as ten *Brahmas*,

(c) and they (ten *Brahmas*) *together* envelop the *Shiv-Lok*,

(d) and all the ten *Brahmas* simultaneously go transcendental.

16. Still further, scriptures are full of descriptions,

(a) of the transcendental phenomena of each of ten *Brahmas* going transcendental simultaneously,

(b) and each *Brahma* acquiring potentialities for creation,

with grace of the lord of transcendental world.

17. With grace of lord of transcendental world,

(a) each of *Brahma* becomes fortunate to have seat of *Lord Shiv*([symbol]) in cavity of his heart,

(b) and each of *Brahma* independently meditates,

(c) and gets multiplied ten fold as ten *Brahmas*

of transcendental potentialities of creation.

18. This transcendental phenomena chase is,
  (a) the transcendental glimpse of Vedic Mathematics,
  (b) which deserves to be chased by the *Sadhkas*,
  (c) and to be self validated,
  (d) as a transcendental glimpse of the transcendental world.

19. Transcendental world,
  (a) unfolding itself,
  (b) as transcendental phenomena,
  (c) for multiplying *Brahmas*,
  (d) as *Mandals* of *Sanatna*,
  (e) as of format of hyper-cube-5,
  (f) as manifestation layer of four folds (with 2-Space(■) in the role of dimension, 3-Space(⊞) in the role of boundary, 4-Space(⌑) in the role of domain and 5-Space(§) in the role of origin) in 4-Space(⌑),

is the,
  (a) unique manifestation feature of the transcendental world,
  (b) with manifestation world itself in the role of origin,
  (c) and also the transcendental world being, the base of the origin,
  (d) and also of the whole range of manifestations remaining within creators space.

20. This transcendental grace,

(a) for creator, the supreme, *Lord Brahma*(B), acquiring powers of creations of whole range of manifested formats with grace of the lord of transcendental world,

(b) and the transcendental world always being lively present at base of each of the manifested formats of creations of creator the supreme, in his own image,

is unique transcendental feature which deserves to be chased by

Sadhkas as transcendental glimpse of *Vedic* mathematics.

21. Creator, the supreme, *Lord Brahma*(⌑),

(a) creating Himself,

(b) as well as, His lord, *Lord Shiv*(§),

(c) and also, the whole range of creations, *Lord Vishnu*(⋈), gods and everybody and everything,

is the unique transcendental feature, which deserves to be chased by the *Sadhkas* as the transcendental glimpse of *Vedic* mathematics.

1.04. VISHWA

22. *Vishwa*,

(a) the world of our existence,

(b) as creation of creator the supreme, *Lord Brahma*(⌑),

is a unique,

(a) transcendental creation of creator,

(b) with grace of lord of transcendental world,

and as such, this creation of *Vishwa deserves* to be,

(a) chased by *Sadhkas*,

(b) as transcendental glimpse of *Vedic* mathematics of creator's creations,

(c) and Vishwa, created in the image of creator with three heads and fourth head as fourth quarter remaining unmanifest.

23. Scriptures are full of descriptions of *VISHWA RUPA* (Form of *Vishwa*/world of our existence)

(a) as existence of Being within frame,

(b) of three heads,

(c) with each head equipped with a pair of eyes.

24. *Sadhkas*,

having,

(a) an urge to know and chase transcendental glimpse of *Vedic* Mathematics of *Vishwa Rupa*,

shall,

(b) go in trans,

(c) and transcend the waking state of consciousness,

(d) and glimpse the linear order of waking state transcendently transforming into a spatial order as three quarters of spatial order,

being sufficient to manifest even the unmanifest fourth quarter of the spatial order in the process.

25. The transcendental glimpse of *Vedic* Mathematics of *Vishwa Rupa*,

(a) as Being within a frame,

(b) with three heads,

(c) and each head equipped with a pair of

eyes,

is the transcendental glimpse,

(i) which deserve to be chased by the *Sadhkas*,

(ii) by transcending waking state of consciousness of linear order

(iii) and glimpsing the manifestation of *Vishwa*,

(iv) as spatial order of the second state of consciousness, as the state of creation within creator's world.

26. The *Sadhkas* shall,

(a) permit the mind to transcend the waking state of consciousness,

(b) to glimpse the dream state of consciousness,

(c) and attain transformation of linear order, into spatial order,

(d) and also the manifestation of fourth un-manifest quarter,

as transcendental glimpse of Vedic Mathematics of the dream state of consciousness.

27. *Sadhkas* shall,

(a) transcend waking state of consciousness time and again,

(b) to perfect intelligence,

(c) to chase transformation of linear order in to spatial order,

(d) and manifestation of un-manifest quarter,

(i) in creator's space,

(ii) as transcendental phenomena,

glimpsed by *Vedic* mathematics,

- as of manifested formats, of four quarters,
- with transcendental base,
- at the origin,
- as the source(E),

for manifestation of four folds.

28. *Sadhkas*

shall go,

(a) in trans,

(b) time and again,

(c) as to chase as to how,

(i) the *Vishwa Rupa,*

(ii) of waking state,

(iii) as of linear order,

- transforms into spatial order,
- and also, gets multiplied four fold,
- with spatial domain getting enveloped within four linear folds,
- and each of the four linear folds going transcendental,
- with spatial order flowing through origins of all the four linear folds enveloping the spatial folds.

29. *Sadhkas,*

shall go,

(a) in trans,

(b) time and again,

(c) to glimpse, the way,

(i) each of the four linear folds, gets potentialities with their origins permitting flow of spatial order

(ii) and there by, each of the four linear folds gets potentialities for further two fold transcendence for spatial order within each of the four linear folds,

(iii) and how with it,
- everything goes transcendental,
- and spatial order flowing through liner folds

transforming these as spatial folds of di-monad formats.

FOR FURTHER CHASE

The list of other Vedic mathematics Books which may be referred for further chase is added as Annexure at the end of present Book.

Maharishi Vedic mathematics (Published in Appendix of the Book 'foundations of higher Vedic mathematics published by M/s Arya Book Depot may be gone through to have complete view of Vedic mathematics domain.

■

ANNEXURE 1

GANITA SUTRAS

01	एकाधिकेन पूर्वेण: *Ekadhiken Purvena*	By One More than One before
02	निखिलं नवतश्चरमं दशत: *Nikhilam Navatascramam Dasatah*	All from 9 and the last from ten
03	ऊर्ध्व तिर्यग्भ्याम् *Urdhva tiryagbhyam*	Vertically and crosswise
04	परावर्त्य योजयेत् *Paravartya Yojayet*	Transpose and Apply
05	शून्य साम्यसमुच्चये *Sunyam Samyasamuccaye*	If the samuccaya is the same it is Zero
06	(आनुरूप्ये) शून्यमन्यत् *(Anurupye) Sunyamanyat*	If one is in Ration the others is Zero
07	संकलनव्यवकलनाभ्याम् *Sankalana-vyavakalanbhyam*	By addition and by subtraction
08	पूरणपूरणाभ्याम् *Puranapuranabhyan*	By the completion or non-completion
09	चलनकलनाभ्याम् *Calana-kalanabhyam*	Differentiation Calculus

10	यावदूनम् *Yavadunam*	By the Deficiency
11	व्यष्टिसमष्टि *Vyastisamastih*	Specific and General
12.	शेषाण्यङ्.केन चरमेण *SesnyankenaCaramena*	The Remainder by the last digit
13	सोपान्त्यद्वयमन्त्यम् *Sopantyadvyamantyam*	The ultimate and twice the penultimate
14	एकन्यूनेन पूर्वेण *Ekanyunena Purvena*	By One less than the One Before
15	गुणितसमुच्चयः *Gunitasamuccayah*	The product of the Sum
16	गुणकसमुच्चयः *Gunaksamuccayah*	All the Multipliers

GANITA UP-SUTRAS

01	आनुरूप्येण **Anurupyena**	Proportionately
02	शिष्यते शेषसंज्ञ *Sisyate Sesasamjnah*	That remains is remainder
03	आद्यमाद्येन अन्त्यमन्त्येन *Adyamadyen Antyamantyena*	First with first Last with last
04	केवलैः सप्तकं गुण्यात् *Kevalaih Saptakam Gunyat*	Only Seven as multiplicand
05	वेष्टनम् *Vestanam*	Osculators
06	यावदूनं तावदूनम् *Yavadunam Tavadunam*	That twice This twice
07	यावदूनम् तावदूनीकृत्य वर्ग च योजयेत् *YavadunamTavadunikrtya VargancaYogayet*	That twice This twice Square and add
08	अन्त्पयोर्दशकेऽपि *Antyayordasake'pi*	Ends to sum as ten
09	अन्तययोरेव *Antyayoreva*	Ends to be in ratio

10	स्मुच्चयगुणितः *Samuccayagunitah*	Samuchya as product
11	लोपन स्थापनाभ्याम् *Lopana Sthapananabhyam*	That missing to be established
12	विलोकनम् *Vilokanam*	By observation
13	गुणितसमुच्चयः समुच्चयगुणितः *GunitaSamuccaya Samuccayagunitah*	Product samuchya Samuchya Product

■

ANNEXURE 2

List of Vedic mathematics Books of the Author

1. Vedic geometry
2. Fermat's last theorem and higher spaces Reality course
3. Foundations of higher Vedic mathematics
4. Goldbach theorem
5. Glimpses of Vedic mathematics

   (**Published by M/s Arya Book Depot, Karol Bagh Delhi**)

6. Learn and teach Vedic mathematics

   ISBN 81-89093-01-9

7. Vedic mathematics decodes SPACE BOOK

   ISBN 81-89093-80-0

8. Practice Vedic mathematics Skills

   ISBN 81-89093-82-7

   (**Published by M/s Lotus press Delhi**)

Vedic mathematics basics Books

Book-1 The teaching of Vedic mathematics